Palgrave Studies in Media and Environmental Communication

Series Editors

Anders Hansen, School of Media, Communication and Sociology, University of Leicester, Leicester, UK
Steve Depoe, McMicken College of Arts and Sciences, University of Cincinnati, Cincinnati, USA

Drawing on both leading and emerging scholars of environmental communication, the Palgrave Studies in Media and Environmental Communication Series features books on the key roles of media and communication processes in relation to a broad range of global as well as national/local environmental issues, crises and disasters. Characteristic of the cross-disciplinary nature of environmental communication, the books showcase a broad variety of theories, methods and perspectives for the study of media and communication processes regarding the environment. Common to these is the endeavour to describe, analyse, understand and explain the centrality of media and communication processes to public and political action on the environment.

Sibo Chen

Petro-Rhetoric in Canada

Narrating Extractivism

palgrave
macmillan

Sibo Chen
School of Professional
Communication
Toronto Metropolitan University
Toronto, ON, Canada

ISSN 2634-6451 ISSN 2634-646X (electronic)
Palgrave Studies in Media and Environmental Communication
ISBN 978-3-032-12963-5 ISBN 978-3-032-12964-2 (eBook)
https://doi.org/10.1007/978-3-032-12964-2

This Palgrave Macmillan imprint is published by the registered company Springer Nature Switzerland AG
The registered company address is: Gewerbestrasse 11, 6330 Cham, Switzerland

Competing Interests The author has no competing interests to declare that are relevant to the content of this manuscript.

Contents

List of Figures

CHAPTER 1

Introduction

Abstract This chapter places Alberta's 2023 energy politics in the context of worsening climate impacts to motivate an inquiry into how state and industry actors legitimize non-conventional fossil fuel development. It traces a discursive shift from outright denial to a diversified persuasive repertoire—economic, cultural, ethical, and techno-optimistic—arguing that contemporary pro-oil sands rhetoric in Canada exemplifies "neoliberal populism," in which market orthodoxy is presented as the people's will. This frames the book's central question: how evolving pro-bitumen narratives normalize continued extraction amid a climate emergency. The chapter concludes with an overview of the public campaigns that will be analysed in subsequent chapters.

Keywords Alberta · Discourse analysis · Climate emergency · Neoliberal populism · Petroculture · Petro-rhetoric · Trans Mountain

In April 2023, the provincial government of Alberta released its updated climate action blueprint. Titled *the Alberta Emissions Reduction and Energy Development Plan* (Government of Alberta, 2023), the document boasted a "made-in-Alberta" approach, which, according to Premier Danielle Smith, would use hydrocarbons "in innovative ways to ensure

S. Chen, *Petro-Rhetoric in Canada*, Palgrave Studies in Media and Environmental Communication,
https://doi.org/10.1007/978-3-032-12964-2_1

Alberta continues to provide the world with sustainably produced energy and products" and reduce emissions "based on reality, instead of unachievable targets set to random timelines" (p. 4).

Considering that the document was released only one month before the 2023 Alberta general election, it was widely viewed as a crucial part of the governing United Conservative Party's (UCP) election platform, whose promises and policy priorities envision fossil fuels remaining a key component of the global energy mix over the coming decades (Graney, 2023). The oil and gas industry was also prominently featured in the election platform of the Alberta New Democratic Party (NDP), the province's official opposition. For example, the Alberta NDP pledged not to implement a provincial retail carbon tax (Graney, 2023). It instead planned to introduce policies that would expand the petrochemical sector and reduce energy project wait times.

The 2023 Alberta general election concluded with the UCP securing a majority government, indicating that fossil fuels would continue to play a significant role in the province's fiscal future for at least the next few years. This result came as no surprise to people familiar with Canadian politics, considering that no other province's economy is as dependent on the energy sector as Alberta's. Nonetheless, the refusal of both the UCP and NDP to reduce Alberta's economic ties with fossil fuels contradicted a glaring climate emergency at hand. At the very moment when both parties were advocating for the centrality of fossil fuels to the provincial economy, the province witnessed an unprecedented occurrence of wildfires during the spring season, wherein a remarkable 520 wildfires ravaged over 1,000,000 hectares (Fortner, 2023). Similarly, in Alberta's neighbour province, British Columbia, the 2023 wildfire season shattered records, with blazes scorching over 22,000 sq. km (i.e., 2,200,000 hectares) (Owen, 2023).

The observed escalation in the severity of wildfires in Western Canada has been linked to climate change (Coogan et al., 2019; Parisien et al., 2023; Tripathy et al., 2023). As many Canadians are already impacted by climate change-fuelled extreme weather events, it is critical to look into the various factors underlying Alberta's insistence that the path it takes to achieve net-zero must be compatible with the oil and gas sector's interests. This book takes on the task by investigating how state and industry actors discursively legitimize non-conventional fossil fuel development (the oil sands sector in the case of Alberta) in an era of climate emergency. As I will detail in subsequent chapters, oil sands advocates have notably transitioned from outright denial of climate change to employing a more diverse array of rhetorical strategies that address issues such as the economy, culture, ethics, and technology. I argue that the evolution of pro-oil sands discourses in Canada since the mid-2010s reflects an increasing embrace of "neoliberal populism," a national-populist mutation of neoliberalism (Bekmen & Alp Özden, 2022; Guardino, 2018; Scheiring, 2022). Central to neoliberal populism are discourses utilizing populist appeals to implement neoliberal economic policies. These discourses mobilize public support by constructing a rigid "us versus them" dichotomy and portraying the policies as being in the best interests of "the people," despite the fact that they tend to disproportionately benefit corporations and the wealthy.

1.1 Alberta Oil Sands: The Controversy

Alberta's oil sands, also referred to as "tar sands" by its opponents, are a mixture of bitumen, clay, sand, and water (Alberta Energy Regulator, n.d.). Located in the province's northern region, the deposits of oil sands are among the world's largest proven unconventional crude oil reserves. The oil sands industry is a major contributor to both the provincial and federal economies, attracting billions of dollars in investment and generating tens of thousands of jobs (Government of Canada, 2016). In recent years, Cenovus Energy, Canadian Natural Resources Limited (CNRL), Imperial Oil Limited, and Suncor Energy have come to dominate oil sands production, "accounting for about 84% of Alberta's daily production of 3.3 million barrels of bitumen" (Arcah, 2022, para. 2). These oil and gas corporations contribute to a complex web of economic, social, and political interests that influence Canada's energy and climate policymaking and pertinent public discourses.

Internationally, the oil sands industry has positioned Canada as a major exporter in the global energy market. This status, however, comes with a glaring caveat. As the Canadian Association of Petroleum Producers (CAPP), the advocacy group representing the upstream Canadian oil and natural gas industry, concedes, "99% of Canada's oil exports currently go to the U.S." (n.d., para. 2). This dependency disadvantages the oil sands industry, which subsequently prompts its desire to "get oil to tidewater" (Harden-Donahue, 2016) by constructing energy infrastructure (pipelines, export ports, etc.) linking Alberta to markets outside North

America. Since the early 2000s, oil sands mega-projects have been at the centre of public debates regarding Canada's energy and climate policies.

The equity and long-term viability of the economic prosperity promised by oil sands are also increasingly under scrutiny. Critics contend that the oil sands industry benefits large corporations and shareholders disproportionately, while local communities and indigenous populations are frequently neglected (Carroll, 2021; Davidson & Gismondi, 2011; Nikiforuk, 2010). For example, a 2018 investigative report by the independent media *Narwhal* (Wilt, 2018) revealed that in 2017, Alberta's 30 most productive oil sands projects, which accounted for more than 95% of oil produced in the province, generated profits of $10.14 billion but only paid $3.21 billion in royalties, taxes, and fees to various governments. Thus, various levels of the Canadian government only received about 32% of the oil sands profits, which is significantly less than what Saudi Arabia (85%), Norway (78%), and China (63.5%) received. The issue of disparity in profit distribution prompts questions about the broader socio-economic impacts of extractive industries in Canada and similar resource-dependent economies.

The oil sands industry's daily operations involve significant water use, land disturbance, and greenhouse gas emissions, and consequently, its environmental impacts continue to be a topic of intense discussion in the Canadian public sphere. According to a Stockholm Environment Institute report (Erickson, 2018), "continued production of the oil sands may lead to an oversupply of oil in global markets, making a low-carbon transition more difficult and undermining global climate action" (p. 2). Likewise, Nikiforuk (2010) views the expanding production of Alberta oil sands as a reckless gamble that not only threatens the local environment but is also unconscionable from a planetary health perspective. In response to these criticisms, the industry has typically emphasized technological solutions (especially carbon capture, utilization, and storage) without addressing larger systemic issues (Carroll, 2021).

Given the divergent perspectives on the socio-economic and environmental implications of oil sands, pertinent political discourses are polarized (Carroll, 2021; Davidson & Gismondi, 2011; Stefanick & Shrivastava, 2015). On the one hand, proponents such as the Alberta government and industry leaders frame oil sands development as a national project integral to Canada's economic strength. On the other hand, opponents, notably environmentalists and many indigenous communities, criticize the continuation of oil sands expansion as emblematic of government shortsightedness and excessive reliance on extractivism. Shedding light on the tensions between the two camps are controversies surrounding mega-energy projects such as the Northern Gateway Pipelines (NG) and the Trans Mountain Pipeline Expansion (TMX). More recently, the industry's responses to the COVID-19 pandemic and low oil prices have prompted public concerns regarding its resilience and long-term viability in a transitioning global energy landscape (Chen, 2023a).

In short, the Alberta oil sands industry, despite its considerable economic contributions at both provincial and federal levels, is confronted by profound controversies and challenges that manifest as a complex interplay of economic, social, political, and environmental discourses. By examining these discourses, this book not only provides a nuanced understanding of the material and discursive structures of energy politics but also raises questions about how extractivism and populism have become deeply interconnected and interdependent.

1.2 Previous Research on Energy Communication

According to Endres et al. (2016), energy communication research is "the study of symbolic practices surrounding material experiences with energy resources, production, and consumption, including related practices of research, development, deployment, and policy" (p. 240). This definition recognizes that communication influences not only how individuals understand energy but also the implications that result from such understandings. Energy communication is primarily defined by the content it investigates, as opposed to the medium and type of communication. Its theoretical and methodological frameworks share many similarities with those of adjacent communication subfields, notably risk and crisis communication and environmental communication.

The majority of energy communication research to date has been conducted within a "crisis frame," which manifests as research examining communicative responses to high-profile energy-related crises as well as broader discourses and frames regarding climate change and humanity's energy futures (Cozen et al., 2018; Endres et al., 2016). For example, extensive research has been conducted on the communicative dynamics of high-profile accidents such as the Fukushima nuclear disaster (e.g., Cotton et al., 2015; Tateno & Yokoyama, 2013) and the BP Deepwater Horizon oil spill (e.g., Breeze, 2012; Chewning, 2015).

Although the crisis frame has advanced an important and productive research agenda, it may be limiting by diverting scholarly attention from questions about the role of energy in daily life (Cozen et al., 2018; Endres et al., 2016). Conceiving energy as an everyday social practice, according to Cozen et al. (2018), "turns attention towards viewing energy transition as a continuous communication phenomenon" (p. 289). This perspective prompts critical examinations of issues such as ordinary people's struggles over energy choice and consumption and everyday practices that form patterned modes of grappling with what energy means. These examinations bear both theoretical and practical implications for promoting a sustainable energy future by changing the way we perceive, communicate, and utilize energy.

Despite a dearth of theorizing energy as a propelling historical force in energy communication, it has served as the starting point for considerable previous research in the burgeoning interdisciplinary field of energy humanities, which pays special attention to the complex interplays between energy systems and contemporary modes of power (Boyer, 2014; Szeman, 2014; Urry, 2014). Energy humanities consider the socio-political and ideological impacts brought by global capitalism's dependency on fossil fuels as a new paradigm of governmentality that renders the management of life and population "increasingly and unstably intertwined with particular infrastructures, magnitudes, and habits of using electricity and fuel" (Boyer, 2014, p. 323).

Political economic analysis is a powerful method for understanding the totality of energy through analysing questions such as how energy makes modernity possible and the underlying structures of capitalism's reliance on fossil fuels. For example, Huber's (2013) analysis of the significance of oil for "American ways of life" suggests that the reproduction of American neoliberalism, which is organized around mobility, individual liberty, property ownership, and entrepreneurship, hinges upon the pivotal yet invisible role of oil as a cheap, accessible, and seemingly inexhaustible energy resource. Similarly, Di Muzio (2015) proposes the idea of "petro-market civilization," arguing that the evolvement of social reproduction is "founded upon non-renewable fossil fuels, mediated by the price mechanism of the market, and dominated by the logic of differential accumulation" (p. 5).

Compared to the productive outcomes of energy humanities in disciplines such as anthropology, cultural studies, geography, and political science, research on how various communications influence public perception of the energy-society relationship has been in short supply. In response to this gap in the literature, the case studies to be reported in subsequent chapters will provide meso-level analyses elucidating how oil sands proponents employ everyday rhetorical tropes to bolster the continuing growth of fossil fuel dependence. The findings will have implications for the three new directions of energy communication outlined by Cozen et al. (2018): (1) activism, social movements, and strategic planning; (2) transnational colonialism and environmental justice; and (3) energy democracy.

Zooming into the Canadian context, research topics regularly explored in previous energy communication scholarship on oil sands include (1) communications surrounding public confrontations over energy infrastructure, (2) the role of news coverage and other public discourses in shaping public opinion and policymaking outcomes, and (3) the discursive struggles over fossil fuel dependence on a daily basis. A major theoretical concern shared by research on these topics is how mediated representations of oil sands contribute to the omnipresence of petro-hegemony (Conway, 2023; McCurdy, 2023). As noted in McCurdy (2023), "media, in myriad forms, offer representational arenas and imaginaries where actors with competing ideas and resources actively engage in a constant barrage of symbolic struggles" (p. 429).

For example, in an early study of how the Alberta government visually branded its resource-based economy, Takach (2013) explained how images promoted by an official place-branding slideshow celebrated extractive capitalism's co-optation of landscapes by juxtaposing themes of opportunity, freedom, and personal achievement with a preponderance of images of pristine, majestic nature. Similarly, studies such as Friedel (2008) and Remillard (2011) have identified tensions between semiotic tropes framing nature as sublime and those framing nature as resources. These tensions contribute to divergent public opinions about resource management and environmental degradation. More recently, lifestyle messaging during oil sands advertising campaigns, which celebrates oil's ubiquity in consumer culture, has received increasing scholarly attention (McCurdy, 2018; Thomlison, 2019). McCurdy's (2018) analysis of change in oil sands advertisements between 2006 and 2015 suggested that industry stakeholders had embraced a "lifestyle turn," as evidenced by visual communications that render the natural environment invisible while reinforcing the notion of oil as the lifeblood of consumer capitalism. The prevalence of the human-made environment in visual promotions of oil sands should be interpreted as part of a larger "climate delay" discourse constructed by the fossil fuel industry and its allies to defend the continuation of petro-based economic and social prosperity.

Concerning media framing of oil sands, public debates sparked by controversies surrounding energy infrastructure have been a major driver for pertinent scholarship. By employing a variety of approaches to media effects, studies such as Chen (2019), Dusyk et al. (2018), Gunster and Saurette (2014), Jekanowski (2018), Raso and Neubauer (2016), Papineau and Deacon (2017), and Way (2011) have contributed to the understanding of pro-oil sands narratives orchestrated by energy corporations and their government and civil allies. These studies have collectively demonstrated how the political and ideological manipulations deployed by oil sands proponents reinforce a "us versus them" dichotomy, which portrays oil sands as indispensable to "Canadian ways of life" while demonizing those who oppose resource dependence. Notably, the discursive construction of "Canadian ways of life" is intentionally ambiguous in order to facilitate the association of the notion with positive connotations. For example, Jekanowski's (2018) analysis of Canadian petro-films from the 1940s and 1950s, which documented oil exploration and development in Western Canada during that era, revealed that the Alberta petroleum industry has, from its inception, strived to normalize its disruptive land-use practices. This is achieved by drawing parallels between petroleum and pre-existing national imaginaries regarding Canada's resource economy.

Meanwhile, mainstream Canadian media discourse has downplayed the disruptive consequences of oil sands development. Papineau and Deacon (2017) noted a geographical disparity in the coverage of oil sands by major Canadian newspapers from 2003 to 2013. During this period, the majority of oil sands coverage was event-specific and predominantly appeared in western Canadian newspapers. This disparity could skew public perceptions of oil sands' ramifications at both local and national levels by downplaying issues such as the global labour and capital flows that underpin Canadian extractive industries and their push for provincial and federal deregulation. The portrayal of oil sands as a regional dispute marked by a trade-off between economic benefit and environmental risk further exacerbates the lack of public awareness about the oil sands industry's large carbon footprint. As found in Dusyk et al.'s (2018) analysis of how media framed NG (which was proposed to carry unrefined oil sands from Alberta to the coast of British Columbia), climate change was mentioned in less than 5% of the articles.

Another trait of oil sands news, as explicated by studies such as Chen (2019) and Gunster and Saurette (2014), is pro-industry media voices' increasingly hostile stance toward environmental critics. In newspapers like the *National Post* and the *Calgary Herald*, pro-oil sands narratives regularly depict the oil sands industry, and to a lesser extent, the people of Alberta, as being aggressively targeted by a well-financed and ideologically motivated global environmental lobby. A key structural element underpinning the orchestrated media attacks against environmentalism is a growing trend of ownership concentration in the Canadian news sector. In particular, the media conglomerate Postmedia Network, whose titles dominate Canadian daily newspaper circulation (Edge, 2016), maintains close ties with the energy sector and is known for its pro-extractivism stance (Neubauer, 2018).

The Postmedia Network's proclivity toward the oil and gas industry exemplifies coordination among members of the Canadian "petro-bloc"—"a decentralized yet interlocked constellation of state, civil society, and corporate actors jointly dedicated to tar sands expansion" (Neubauer, 2018, p. 249). The Gramscian concept of a (historical) bloc underscores the central role of hegemony across various social fields, such as the state, civil society, and media, in sustaining popular support for oil sands expansion. According to previous research (Neubauer, 2018; Raso & Neubauer, 2016), conservative think tanks, foundations, and advocacy groups perform a pivotal function in establishing petro-hegemony. Their communications, disseminated through pro-industry media and other public channels, lend coherence and validity to the priorities set by the oil and gas industry and their state allies. Further amplifying such communications is a subsidized petro-public mainly comprised of people who have close ties with the oil and gas industry (Wood, 2018). The active participation of the petro-public in oil sands promotional campaigns, such as "Canada's Energy Citizens," indicates a concerning trend, namely a fusion of corporate public relations and citizen political participation.

Discourses promoting fossil fuels, orchestrated by the Canadian petro-bloc, face intense opposition, notably from progressive think tanks, environmental organizations, Indigenous communities, and concerned citizen groups. Studies on these groups' discursive resistance to the oil and gas industry primarily draw from the literature of social movements, focusing on how they transcend the conventional "jobs versus the environment" dichotomy. Chen's (2023a, 2023b) analysis of the controversies surrounding the proposed establishment of a liquefied natural gas industry in British Columbia, for instance, detailed how environmentalists and their allies worked in concert to raise public awareness about the implications of these controversies for the integrity of democracy, local governance, and Indigenous sovereignty. These political issues also constitute the major concerns of those opposing Alberta oil sands, whose collective actions, as noted in previous research such as Belliveau et al. (2021) and Shaw (2021), indicate a push toward "energy democracy"—a vision combining energy decarbonization with economic democratization.

Social license is a noteworthy concept that has been politicized in the context of energy infrastructure conflicts, serving to underscore the importance of energy democracy. According to Gunster and Neubauer (2019), the concept, once championed by the oil and gas industry to win broad public support for its projects, "has come to symbolize the failure of corporations as well as governments to secure legitimacy and public approval for specific projects, as well as an opportunity to assert the radical, democratic autonomy of local communities (especially First Nations) to reject certain forms of development as fundamentally incompatible with their needs, interests, and values" (p. 708). This is because the concept's implicit legitimation of the capacity of social actors to reject or withhold a "license" enables critics of the oil and gas industry to appropriate it to draw public attention to the unequal distribution of benefits and lack of democratic accountability that characterize Canada's extractive industries. Moreover, as what is considered a legitimate means to revoke a social license is a subject of contention, many activists justify direct action in land defence by invoking the concept, as evidenced by the worldwide proliferation of blockadia movements (Carroll, 2021; Chen, 2021).

In sum, this book contributes to the growing body of environmental communication research by providing an in-depth analysis and critical assessment of a key player in global energy transition: Canada. Its investigation into the recent developments of discursive politics of oil sands maps elucidates the complexity underlying diverse stakeholders' economic, political, social, and cultural encounters with the abstract notion of resource-driven prosperity. By doing so, the book demonstrates that to gain a comprehensive understanding of the difficulties linked to challenging the hegemony of petro-dependency, close attention must be paid to the narrative construction and discursive symbolization processes.

1.3 Methods and Organization

Two overarching objectives guided the analysis reported in the subsequent chapters: (1) to elucidate how the Canadian petro-bloc constructs discourses to garner public support for the continuing expansion of oil sands, and (2) to assess how these discourses reinforce specific interpretations of oil sands while marginalizing or excluding alternative perspectives. I achieved both objectives by examining three prominent public campaigns and their related textual materials (websites, press releases, news articles, etc.) concerning two contentious mega oil sands projects: NG and TMX. These public campaigns were analysed following the discourse-historical approach (DHA), a variant of critical discourse studies (CDS).

In line with Wodak and Meyer (2016), I consider CDS an umbrella term referring to critical methods that expose and problematize commonly held beliefs regarding language, power, and society, while also acknowledging discourse as a potentially influential agent for change. DHA distinguishes itself from other CDS methods through its emphasis on interdisciplinarity, drawing insights from various disciplines; historical context, recognizing discourses as part of broader historical patterns and developments; and data triangulation, employing a wide range of methods to gather and analyse data (Reisigl, 2017; Reisigl & Wodak, 2016). In practice, DHA-inspired studies have investigated the discourses surrounding a vast array of subjects, including but not limited to discrimination, history, identity, and media representation. Their analytical frameworks typically centre on the examination of a text's linguistic nuances in conjunction with its historical background, with the intention of comprehending its meaning and impact on society. Reisigl (2017) suggests five essential queries (Table 1.1) that DHA utilizes to examine the discursive features of a text.

Inspired by the queries in Table 1.1, my analysis aimed to identify macro-patterns shared by multiple texts as well as link these patterns to the context in which the texts emerge and are used. In consultation with Anna Kuteleva's (2021; Kuteleva & Leifso, 2020) previous research on identities and historical narratives about energy resources, my analysis focused on the intertextuality between discourses constructed by different stakeholders. Four interconnected definitional queries (Kuteleva & Leifso, 2020) were posed during my reading and coding of the collected data: (1) What is oil sands? (2) Who are "we" in relation to oil sands? (3) Who

Table 1.1 Discursive Strategies in the Discourse-Historical Approach (Reisigl, 2017, p. 52)

Questions to approach discursive features	*Discursive strategies*	*Purpose*
How are persons, objects, phenomena, events, processes and actions named and referred to linguistically in the discourse in question?	Nomination	Discursive construction of social actors; discursive construction of objects, phenomena, events; discursive construction of processes and actions
What characteristics or qualities are attributed to social actors, objects, phenomena, events, processes, and actions mentioned in the discourse?	Predication	Discursive characterization of social actors, objects, phenomena, events, processes and actions (e.g., positively or negatively)
What arguments are employed in discourse?	Argumentation	Persuading addressees of the validity of specific claims of truth and normative rightness
From what perspective are these nominations, attributions, and arguments expressed?	Perspectivization	Positioning the speaker's or writer's point of view and expressing involvement or distance
Are the respective utterances articulated overtly? Are they intensified or mitigated?	Mitigation and intensification	Modifying the illocutionary force of utterances in respect to their epistemic or deontic status

are "they" in relation to oil sands? (4) What are "we" going to do with oil sands?

The rest of the book proceeds in four chapters. Each chapter commences with a series of theoretical arguments drawn from a synthesis of previous research, followed by a qualitative case study of a representative, pro-oil sands public campaign and its principal rhetorical appeals established by the Canadian petro-bloc. I then discuss the wider public responses (e.g., reports and opinion pieces in mainstream and alternative media) provoked by the public campaign, thereby unveiling various stakeholders' support for and opposition to the oil sands industry.

Chapter 2 attends to the oil sands industry's shift toward lifestyle messaging, which legitimizes the further expansion of fossil fuel extraction by celebrating oil's pervasiveness in consumer culture. By examining the discursive strategies found in "Life Takes Energy," a massive public campaign launched by Enbridge to promote NG in 2014, the chapter elucidates how the rhetoric of "petro-prosperity" is constructed via narratives around ordinary moments of everyday life and how these moments, ultimately, rely upon the operation of energy infrastructure. As such, "Life Takes Energy" and similar campaigns launched elsewhere can be considered as manifestations of petroculture, in which fossil fuel dependency functions as an epistemological justification confining our socio-cultural imaginary.

Chapter 3 examines the political contest over TMX between 2015 and 2018. During this period, Alberta was governed by the provincial New Democratic Party, which adopted a more progressive stance in safeguarding the interests of the oil sands industry in comparison with the provincial Conservative Party that governed the province for over four decades (1971–2015). This chapter provides a critical analysis of "Keep Canada Working," a short-lived campaign launched by the Alberta NDP government in 2018 in an attempt to gain national support for TMX. The analysis focuses on the rhetoric of "symbolic nationalization," which, by concealing the fact that the oil sands industry is a private sector heavily controlled by foreign capital, seeks to define extractivism as part of the Canadian identity. The chapter ends by highlighting the glaring contradiction between what homogenizing populist appeals promise and the reality on the ground.

In Chapter 4, the TMX controversy is further examined with a focus on its developments since the Conservatives were re-elected in the 2019 Alberta general election. The current Conservative government of Alberta has shifted the emphasis of its pro-bitumen narratives to portray opponents of the bitumen industry as radicals supported by foreign interests. This chapter analyses this turn to smear tactics by linking it to the rise of "extractive populism" in Canada. I argue that this trend reflects the growing radicalization of Canada's petro-bloc, whose members embrace alternative facts to deny the broad public consensus on the phase-out of fossil fuels. This argument is developed through a critical analysis of discourses associated with the "Public Inquiry into Anti-Alberta Energy Campaigns."

Chapter 5 begins with a summary of the main findings from the case studies and the key takeaways regarding the pitfalls of pro-oil sands discourses. Recognizing the immediate need for public mobilization toward decarbonization, the chapter then discusses strategies to mitigate the polarization currently prevailing in public debates about Canada's energy future. It concludes by emphasizing the necessity of further work to dispel the "common-sense" beliefs the public associates with fossil fuels, particularly given the prevalence of petroculture in public discourse.

References

Alberta Energy Regulator. (n.d.). *Oil sands*. https://www.aer.ca/providing-information/by-topic/oil-sands

Ascah, R. L. (2022, August 21). The Big Four oilsands companies' influence threatens Alberta democracy, argues political scientist. *Conversation Canada*. https://theconversation.com/the-big-four-oilsands-companies-influence-threatens-alberta-democracy-argues-political-scientist-188567

Bekmen, A., & Alp Özden, B. (2022). The rise and demise of neoliberal populism as a hegemonic project: Brazil, Thailand, and Turkey. *Rethinking Marxism, 34*(3), 338–360. https://doi.org/10.1080/08935696.2022.2069459

Belliveau, E., Rowe, J. K., & Dempsey, J. (2021). Fossil fuel divestment, non-reformist reforms, and anti-capitalist strategy. In W. K. Carroll (Ed.), *Regime of obstruction: How corporate power blocks energy democracy* (pp. 453–478). Athabasca University Press.

Boyer, D. (2014). Energopower: An introduction. *Anthropological Quarterly, 87*(2), 309–333. https://doi.org/10.1353/anq.2014.0020

Breeze, R. (2012). Legitimation in corporate discourse: Oil corporations after Deepwater Horizon. *Discourse & Society, 23*(1), 3–18. https://doi.org/10.1177/0957926511431511

Canadian Association of Petroleum Producers. (n.d.). *Canada's energy mix*. https://www.capp.ca/energy/canadas-energy-mix/

Carroll, W. K. (ed.) (2021). *Regime of obstruction: How corporate power blocks energy democracy*. Athabasca University Press.

Chen (2019), From persuasion to manipulation: Tracing Oil Sands narratives in the *Calgary Herald*. In H. Graves & D. E. Beard (Eds.), *The rhetoric of oil in the twenty-first century: Government, corporate, and activist discourses* (pp. 255–270). Routledge.

Chen, S. (2021). The rise of blockadia as a global anti-extractivism movement. *Local Environment, 26*(12), 1423–1428. https://doi.org/10.1080/13549839.2021.1969352

Chen, S. (2023a). The fate of bitumen: An exploratory study of national newspaper coverage of Alberta's bitumen industry during the COVID-19 pandemic. *Humanities & Social Sciences Communications, 10*, Article 30. https://doi.org/10.1057/s41599-023-01515-2

Chen, S. (2023b). *Energy politics and discourse in Canada: Probing progressive extractivism*. Routledge.

Chewning, L. V. (2015). Multiple voices and multiple media: Co-constructing BP's crisis response. *Public Relations Review, 41*(1). https://doi.org/10.1016/j.pubrev.2014.10.012

Coogan, S. C. P., Robinne, F.-N., Jain, P., & Flannigan, M. D. (2019). Scientists' warning on wildfire — a Canadian perspective. *Canadian Journal of Forest Research, 49*(9), 1015–1023. https://doi.org/10.1139/cjfr-2019-0094

Conway, K. (2023). Reading oil (back) into media history: The case of postwar television. *International Journal of Cultural Studies, 26*(4), 410–426. https://doi.org/10.1177/13678779221129295

Cotton, A. J., Veil, S. R., & Iannarino, N. T. (2015). Contaminated communication: TEPCO and organizational renewal at the Fukushima Daiichi nuclear power plant. *Communication Studies, 66*(1), 27–44. https://doi.org/10.1080/10510974.2013.811427

Davidson, D. J., & Gismondi, M. (2011). *Challenging legitimacy at the precipice of energy calamity*. Springer Press.

Di Muzio, T. (2015). *Carbon capitalism: Energy, social reproduction and world order*. Rowman & Littlefield.

Dusyk, N., Axsen, J., & Dullemond, K. (2018). Who cares about climate change? The mass media and socio-political acceptance of Canada's oil sands and northern gateway pipeline. *Energy Research & Social Science, 37*, 12–21. https://doi.org/10.1016/j.erss.2017.07.005

Edge, M. (2016). The never-ending story: Postmedia, the competition bureau, and press ownership concentration in Canada. *Canadian Journal of Media Studies, Spring/summer, 2016*, 53–81.

Erickson, P. (2018). Confronting carbon lock-in: Canada's oil sands. *SEI discussion brief*. Stockholm Environment Institute. https://www.sei.org/publications/confronting-carbon-lock-canadas-oil-sands/

Endres, D. E., Cozen, B., Trey Barnett, J., O'Byrne, M., & Rai Peterson, T. (2016). Communicating energy in a climate (of) crisis. *Annals of the International Communication Association, 40*(1), 419–447. https://doi.org/10.1080/23808985.2015.11735267

Friedel, T. L. (2008). (Not so) crude text and images: Staging native in 'big oil' advertising. *Visual Studies, 23*(3), 238–254. https://doi.org/10.1080/14725860802489908

Fortner, C. (2023 May 23). *2023 wildfires in Alberta worst on record for spring*. CityNews Edmonton. https://edmonton.citynews.ca/2023/05/23/worst-spring-wildfire-season/

Government of Canada. (2016). *Oil sands: Economic contributions*. https://natural-resources.canada.ca/energy/publications/18756

Government of Alberta. (2023, April 19). *The Alberta emissions reduction and energy development plan*. https://open.alberta.ca/publications/alberta-emissions-reduction-and-energy-development-plan

Graney, E. (2023, May 23). Breaking down UCP and NDP energy policies in Alberta. *The Globe and Mail*. https://www.theglobeandmail.com/canada/alberta/article-breaking-down-ucp-and-ndp-energy-policies-in-alberta/

Guardino, M. (2018). Neoliberal populism as hegemony: A historical-ideological analysis of US economic policy discourse. *Critical Discourse Studies, 15*(5), 444–462. https://doi.org/10.1080/17405904.2018.1442361

Gunster, S., & Neubauer, R. J. (2019). (De)legitimating extractivism: The shifting politics of social licence. *Environmental Politics, 28*(4), 707–726. https://doi.org/10.1080/09644016.2018.1507290

Gunster, S., & Saurette, P. (2014). Storylines in the sands: News, narrative, and ideology in the *Calgary Herald*. *Canadian Journal of Communication, 39*(3), 333–359. https://doi.org/10.22230/cjc.2014v39n3a2830

Harden-Donahue, A. (2016, April 18). Why the oil to tidewater argument for pipelines is bunk. *The Narwhal*. https://thenarwhal.ca/why-oil-tidewater-argument-pipelines-bunk/

Huber, M. T. (2013). *Lifeblood: Oil, freedom, and the forces of capital*. University of Minnesota Press.

Jekanowski, R. W. (2018). Fuelling the nation: Imaginaries of western oil in Canadian nontheatrical film. *Canadian Journal of Communication, 43*(1), 111–125. https://doi.org/10.22230/cjc.2017v43n1a3290

Kuteleva, A. (2021). *China's energy security and relations with petrostates: Oil as an idea*. Routledge.

Kuteleva, A., & Leifso, J. (2020). Contested crude: Multiscalar identities, conflicting discourses, and narratives of oil production in Canada. *Energy Research & Social Science, 70*, Article 101672. https://doi.org/10.1016/j.erss.2020.101672

McCurdy, P. (2018). From the natural to the manmade environment: The shifting advertising practices of Canada's oil sands industry. *Canadian Journal of Communication, 43*(1), 33–52. https://doi.org/10.22230/cjc.2018v43n1a3315

McCurdy, P. (2023). The rise and fall of the Synthetic: The mediatization of Canada's oil sands. *International Journal of Cultural Studies, 26*(4), 427–444. https://doi.org/10.1177/13678779231159697

Neubauer, R. (2018). Moving beyond the petrostate: Northern gateway, extractivism, and the Canadian petrobloc. *Studies in Political Economy, 99*(3), 246–267. https://doi.org/10.1080/07078552.2018.1536369

Nikiforuk, A. (2010). *Tar sands: Dirty oil and the future of a continent* (Revised ed.). Greystone Books.

Owen, B. (2023 September 10). B.C.'s wildfire crisis arrived decades earlier than experts forecast. *CBC News*. https://www.cbc.ca/news/canada/british-columbia/bc-wildfires-climate-2023-1.6962307

Papineau, J. W., & Deacon, L. (2017). Fort McMurray and the Canadian oil sands: Local coverage of national importance. *Environmental Communication, 11*(5), 593–608. https://doi.org/10.1080/17524032.2017.1289107

Parisien, M. -A., Barber, Q. E., Bourbonnais, M. L., Daniels, L. D., Flannigan, M. D., Gray, R. W., Hoffman, K. M., Jain, P., Stephens, S. L., Taylor, S. W., & Whitman, E. (2023). Abrupt, climate-induced increase in wildfires in British Columbia since the mid-2000s. *Nature Communications Earth & Environment, 4*, Article 309. https://doi.org/10.1038/s43247-023-00977-1

Raso, K., & Neubauer, R. J. (2016). Managing dissent: Energy pipelines and 'new right' politics in Canada. *Canadian Journal of Communication, 41*(1), 115–133. https://doi.org/10.22230/cjc.2016v41n1a2777

Reisigl, M. (2017). The discourse-historical approach. In J. Flowerdew, J. E. Richardson (Eds.), *The Routledge handbook of critical discourse studies* (p. 44–59). Routledge. https://www.routledgehandbooks.com/doi/10.4324/9781315739342.ch3

Reisigl, M., & Wodak, R. (2016). The discourse-historical approach (DHA). In R. Wodak & M. Meyer (Eds.), *Methods of critical discourse studies* (3rd ed., pp. 23–61). Sage.

Remillard, C. (2011). Picturing environmental risk: The Canadian oil sands and the national geographic. *International Communication Gazette, 73*(1–2), 127–143. https://doi.org/10.1177/1748048510386745

Scheiring, G. (2022). The national-populist mutation of neoliberalism in dependent economies: The case of Viktor Orbán's Hungary. *Socio-Economic Review, 20*(4), 1597–1623. https://doi.org/10.1093/ser/mwac007

Shaw, K. (2021). Flashpoints of possibility: What resistance reveals about pathways toward energy transition. In W. K. Carroll (Ed.), *Regime of obstruction: How corporate power blocks energy democracy* (pp. 399–428). Athabasca University Press.

Stefanick, L., & Shrivastava, M. (eds.) (2015). *Alberta oil and the decline of democracy in Canada*. Athabasca University Press.

Szeman, I. (2014). Conclusion: On energopolitics. *Anthropological Quarterly, 87*(2), 453–464. https://doi.org/10.1353/anq.2014.0019

Takach, G. (2013). Selling nature in a resource-based economy: Romantic/extractive gazes and Alberta's bituminous sands. *Environmental Communication, 7*(2), Article 211230. https://doi.org/10.1080/17524032.2013.778208

Tateno, S., & Yokoyama, H. M. (2013). Public anxiety, trust, and the role of mediators in communicating risk of exposure to low dose radiation after the Fukushima Daiichi Nuclear Plant explosion. *Journal of Science Communication, 12*(2), 1–22. https://doi.org/10.22323/2.12020203

Thomlison, A. (2019). From pipeline to plate: The domestication of oil sands through visual food analogies. In H. Greaves & D. Beard (Eds.), *The rhetoric of oil in the twenty-first century: Government, corporate, and activist discourses* (pp. 189–210). Routledge.

Tripathy, K. P., Mukherjee, S., Mishra, A. K., Man, M. E., & Williams, A. P. (2023). Climate change will accelerate the high-end risk of compound drought and heatwave events. *Proceedings of the National Academy of Sciences, 120*(28), Article e2219825120. https://doi.org/10.1073/pnas.2219825120

Urry, J. (2014). The problem of energy. *Theory, Culture & Society, 31*(5), 3–20. https://doi.org/10.1177/0263276414536747

Way, L. (2011). An energy superpower or a super sales pitch? Building the case through an examination of Canadian newspapers coverage of oil sands. *Canadian Political Science Review, 5*(1), 74–98.

Wilt, J. (2018, August 14). Are Albertans collecting a fair share of oilsands wealth? *The Narwhal.* https://thenarwhal.ca/are-albertans-collecting-a-fair-share-of-oilsands-wealth/

Wodak, R., & Meyer, M. (Eds.) (2016). *Methods of critical discourse studies* (3rd ed.). sage.

Wood, T. (2018). Energy's citizens: The making of a Canadian Petro-public. *Canadian Journal of Communication, 43*(1), 75–92. https://doi.org/10.22230/cjc.2017v43n1a3312

CHAPTER 2

"Life Takes Energy": Promoting Carbon Dependency via Lifestyle Advertising

Abstract Through critical multimodal discourse analysis of Enbridge's "Life Takes Energy" (2014), the chapter explicates how lifestyle advertising rebrands hydrocarbons as the intimate infrastructure of everyday life. By saturating "energy" with scenes of family, care, and convenience, the campaign backgrounded extraction and pipelines and stabilized a Gramscian common sense of benign, inevitable carbon dependence—"selling oil sands without oil sands." The chapter theorizes this as petroculture's affective pedagogy that converts consumption into consent for continued fossil expansion.

Keywords Consent/legitimation · Carbon dependency · Enbridge · Lifestyle advertising · Multimodality · Petroculture

This chapter analyses the oil sands industry's strategic foray into lifestyle advertising, which in recent years has emerged as a prevalent tactic to shape public consciousness about energy-society relations (Cozen, 2010; McCurdy, 2018; Thomlison, 2019). The case under scrutiny is Enbridge's "Life Takes Energy" campaign, which, unveiled in 2014, wove fossil fuels into the fabric of everyday life. By deploying a wide range of multimodal discursive constructions, the campaign portrayed oil and

S. Chen, *Petro-Rhetoric in Canada*, Palgrave Studies in Media and Environmental Communication,
https://doi.org/10.1007/978-3-032-12964-2_2

gas transported by Enbridge as the lifeblood of consumer culture, energizing positive elements of life such as communal bonds, familial warmth, and happiness. The celebration of such positive connotations effectively obscured the negative impacts associated with the extraction, transportation, and consumption of fossil fuels. In short, lifestyle advertising not only enables "selling oil sands without oil sands" (McCurdy, 2018), but also solidifies a fundamental message of petroculture: the world is inextricably constructed and sustained by fossil fuels, rendering their phase-out from our social and political imaginaries a seemingly insurmountable challenge.

2.1 Advertising and Petroculture

McCurdy (2023) argues that "media, in myriad forms, offer representational arenas and imaginaries where actors with competing ideas and resources actively engage in a constant barrage of symbolic struggles" (p. 429). These symbolic struggles manifest as conflicting narratives found in a wide range of media texts, such as advertisements, films, news reports, and TV programmes. In recent years, social media platforms have emerged as major battlegrounds that amplify the contention present in these media texts. Accordingly, while the analytical focus of this chapter is on oil sands advertisements, it should be kept in mind that they are only a small segment of the resource's media and cultural representations that influence how the public makes sense of the social, cultural, political, and environmental issues tied to Canada's oil sands industry.

This chapter considers oil sands advertisements as key contributors to petro-hegemony, which LeQuesne (2019) defines as a matrix of dominance exploited by the fossil fuel industry to manipulate three different relations of power: consent, coercion, and compliance. Advertisements that either extol the economic and societal benefits of the industry or attack its adversaries cultivate a Gramscian common sense concerning the indispensability of fossil fuels to society, thereby subjecting themselves to protracted public censure and scholarly scrutiny.

In 2010, the environmental organization Sierra Club of Canada lodged a complaint with Advertising Standards Canada alleging that an advertisement by the Canadian Association of Petroleum Producers (CAPP) misled the public by comparing oil sands tailings (the toxic by-product of oil sands production) to yogurt (Simpson, 2010). In the opinion of the

Sierra Club of Canada, this analogy violated the Canadian Code of Advertising Standards by downplaying the toxicity of tailings ponds, thereby misleading the public.

Similarly, in the United States, advertising campaigns such as Shell's "Real Energy" and Chevron's "Human Energy" sparked controversy due to their use of food imagery to justify the necessity of expanding fossil fuel production (Cozen, 2010). The advertisements of both campaigns encouraged viewers to consider that all humans need food for survival, while the companies offer a comparable source of energy essential for economic development.

Why is food imagery such a prevalent rhetorical trope in campaigns promoting fossil fuels? This, according to Cozen (2010), is due to a material linkage between fossil fuel energy and human physical energy: just as food provides nourishment for the body, fossil fuel companies assert that they supply the content necessary for society to function. This linkage supports an ideological claim that, by invoking food imagery, naturalizes the perpetual expansion of energy production in two ways. First, energy production is placed on the same discursive plane as food production. Second, food production is presented as an input to the economic system—like fuel for the body—thereby framing both as natural, necessary components of social life. From this framing follows the argument that expanding energy development is a "natural" response to economic growth and population increase, analogous to expanding food production to meet rising bodily needs.

Echoing Cozen's view, Thomlison (2019) considers the repeated presence of visual and textual references to food in oil sands advertisements as energy corporations' deployment of the rhetorical strategy "domestication." Domestication frames large-scale, contested industrial activities in the imagery and language of everyday life—home, family, meals, and routine care—so that these activities appear familiar, necessary, and morally acceptable. By performing such rhetorical work, oil sands advertisements exploit food's strong ties to family and daily life to evoke a feeling of authenticity and positivity in the audience, with the principal objective of garnering social approval for the continuous growth of oil sands production.

However, the rhetorical manoeuvre of drawing parallels between the indispensability and desirability of food and fossil fuels signifies a capitalist appropriation that rationalizes perpetual production and consumption expansion (Thomlison, 2019). Linking oil production to the ideas of household comfort and nourishment signified by food imagery reduces the perceived distance and complexity, while emphasizing personal stories and convivial scenes. Yet, it also obscures environmental harms, Indigenous land claims, and systemic power relations. This leads to a depoliticized, normalized representation that promotes public acceptance and safeguards the legitimacy of corporations.

In addition to the proliferation of food imagery, advertising campaigns that extol the indispensability of fossil fuels in sustaining everyday life have increased significantly in recent years. According to McCurdy's (2018) analysis of print advertisements produced by the oil sands industry from 2006 to 2015, in Canada 2011 marked the start of a lifestyle angle, and then the oil sands industry's efforts became particularly pronounced in subsequent years, with more than 50% of advertisements in 2014 featuring lifestyle issues, including Enbridge's "Life Takes Energy" campaign, which will be analysed later in this chapter. This trend, according to McCurdy, marked a major shift in industry tactics, away from attempting to appease environmental critics and toward implementing a multifaceted communication plan in which the industry defends itself on its own terms. Instead of solely focusing on technical rebuttals and regulatory compliance, the oil sands industry began to reposition and reframe extractive activity as integral and even beneficial to everyday routines and aspirations.

Central to the "lifestyle turn" is the idea of "advertising the oil sands without the oil sands." In an attempt to mitigate the adversity caused by the public impression of "dirty oil sands," corporations such as Enbridge, Cenovus, and Suncor deployed lifestyle advertisements to rebrand themselves as energy providers. In these advertisements, natural environments associated with oil sands production (e.g., landscapes scarred by industrial infrastructure) are replaced by domestic settings (e.g., cooking, commuting, childcare, leisure) that illustrate everyday life. By situating messages within the register of lifestyle advertising, these advertisements leverage the affective power of everyday life to foster emotional resonance and trust. They also render fossil fuel consumption morally neutral or even benevolent by presenting it as ubiquitous yet invisible.

The transition from the natural to the domestic environment, although it may seem familiar and comforting, has political implications. Along with other promotional messages from the oil sands industry, it must be seen as part of a continuous effort to persuade us that we need energy products obtained from the oil sands. Emphasizing the domestic environment reframes debates about the environmental impact of fossil fuels, presenting contemporary society's fossil fuel dependency as a trade-off for progress and convenience rather than as the result of systemic choices about energy futures. Considering the climate emergency, however, it is crucial for our survival that we gradually reduce our dependence on such products.

The decline of natural environment-related representations in oil sands advertisements does not mean that the industry has forsaken environmental communication. It continues to showcase its environmental stewardship through dedicated websites, corporate social responsibility reports, and other promotional materials. This presents an intriguing contrast: while the industry has not diversified its core business, it has notably diversified its approaches to convincing the public of the benefits and costs of oil sands. For example, in a critical visual discourse analysis of an official place-branding slideshow produced by the provincial government of Alberta, Takach (2013) identified two types of "gazes" in the slideshow's visual narratives: the "romantic gaze," which idealizes nature, and the "extractive gaze," which treats nature merely as a resource. Both gazes were used to justify subordinating natural resources to human interests.

The prevalence of lifestyle advertising, which seeks to construct the "Energy = Prosperity" equation, also demonstrates the industry's ambition to set the public agenda on the energy transition. By linking energy production and consumption directly to comfortable homes, good jobs, and community well-being, images and stories repeatedly associate everyday prosperity with fossil fuels. Consequently, public discussions concerning energy choices become cultural debates about identity and livelihood.

Contemporary culture is referred to as "petroculture" by Szeman (2017), who urges research into the ways in which petroleum "narrates, shapes, and circumscribes" our daily lives, relationships, cultures, and imaginations. Meanwhile, communication scholars develop the notion of mediatization to theorize how our social relations, practices, institutions, and physical environment are underpinned and (re)constructed by a pervasive and ever-evolving media and communication environment (Hjarvard, 2008; McCurdy, 2023). Mediatization is considered as a metaprocess that influences our material and mediated lives, in conjunction with other meta-processes, including globalization and commercialization.

Thus, as mediated representations, oil sands advertisements influence the public's understanding of not only the industry but also petroculture more generally. In the case of lifestyle advertisements, their "selling oil sands without oil sands" approach endeavours to replace the prevailing negative perceptions of oil sands with the abstract concept of energy and its positive connotations. This helps oil hide in plain sight in cultural representation, similar to how pipelines mask the negative social and political impact of oil transportation. Besides the material function of getting energy from extraction sites in the hinterland to the postmodern spaces where energy is consumed, pipelines play a crucial ideological role, as noted by petroculture scholars (e.g., Barney, 2017; Macdonald, 2017), in concealing the adverse impacts of oil extraction and transportation from the public, which aids in the continued operation of the oil and gas industry amid mounting concerns about climate change.

Another perspective to analyse the ideological impact of advertisements associated with the oil and gas industry is by examining their contribution to the promotion of privatism in politics. In his incisive analysis of the United States' dependence on fossil fuels, Huber (2013) convincingly argues that what makes "oil addiction" intrinsic to the American public's vision of "a good life" is oil consumption's broader implications for social relations. While the physical properties of oil (e.g., liquidity, versatility, and abundance) and the financial and political power of energy conglomerates are crucial in establishing oil as the primary energy source for capitalism reproduction, the patterns and magnitude of energy consumption, as well as the resulting social and political implications, are equally important in defining oil's current position in the world.

According to Huber (2013), the New Deal era marked a critical juncture in history, during which the United States' oil consumption assumed its present-day dominance. By popularizing increasing wages and single-family suburban lifestyles, the era fostered a life imagery centred around mass consumption of material and space. Although petroleum-powered personal transportation entered the popular imagination of middle-class lifestyles as early as the 1920s, it was not until the New Deal era that automobility became an ingrained component of the "American way of life."

The significance of driving to Americans is evident in the emergence of automobile culture (Paterson, 2007), which is reflected in both popular culture products and advertisements. In addition to automobility, oil consumption underlies two other pillars of the American way of life: suburbia and consumerism. These pillars result in a geographically dispersed social order, underlying which are the oil industry and its technological, geographical, and regulatory systems. The hallmarks of this social order are single-family homes, cars, and gadgets, whose prevalence in ordinary households is facilitated by advertisements. Accordingly, advertisements furnish consumers with the discourses and fantasies that facilitate the mapping of the American way of life onto their daily lives.

Consider, for example, lifestyle advertisements depicting picturesque suburban streets, children being transported to school, couples embarking on weekend excursions, or professionals commuting in comfort—visuals that interweave freedom, familial obligations, leisure, and responsibility with driving. Film and television perpetuate the same script: protagonists gain intimacy, autonomy, status, and personal transformation while driving. Even ordinary service advertisements (e.g., insurance, coffee, fast food, etc.) frequently use automobile-centric scenarios. Taken together, these representations do more than reflect existing daily life. They establish extensive car use as the default setting of modern life and instruct the public to view the built environment through automotive logic, while in tandem dismissing alternatives—such as walking, cycling, and public transit—as inconvenient, marginal, or indicative of a different social identity.

Meanwhile, critiques of the hypocrisy of public figures and institutions—those who openly endorse climate action while privately maintaining high-carbon lifestyles (such as driving luxury cars)—have been increasingly co opted by the oil and gas industry as defensive rhetoric to defend continued reliance on fossil fuels. By citing such criticisms as proof that climate politics are performative, inconsistent, and ultimately pointless, this strategy cultivates public cynicism regarding the viability of deep decarbonization and shifts the focus from institutional accountability to personal moral failure.

Overall, the American public considers oil to be indispensable for their day-to-day activities. Deriving from this view, they establish a conceptual linkage between oil and personal freedom. Advertisements reinforce the linkage by chanting neoliberal concepts of fair competition, freedom, and entrepreneurship. The linkage is similarly evident in Canada, where communities, with the exception of a handful of dense metropolitan cores (Toronto, Montréal, Vancouver, and parts of Ottawa and Calgary), are geographically dispersed, and driving is an essential component of daily life. Much of Canada, like the US, is structured around low-density settlement patterns, such as satellite towns, suburban subdivisions, exurban bedroom communities, and large, sprawling cities connected by regional highways. Postwar housing policy, highway investments, and land-use regulations that prioritized single-family lots, detached homes, and easy access to the highway over compact, mixed-use development all influenced this built environment. The most practical choice for work, school, shopping, and social life is to own a car and drive it daily. Therefore, I anticipate that the "Life Takes Energy" campaign will emphasize the role of driving in facilitating "happy middle-class lifestyles." Visual narratives will be employed to justify the continued use of fossil fuels as pragmatic compromises and to advocate for technological solutions and incrementalism in dealing with the climate crisis.

2.2 The Northern Gateway Pipelines Controversy

In 2008, Enbridge, a Calgary-based pipeline and energy company, proposed the Northern Gateway Pipelines project (NG), a twin-pipeline system that would run approximately 1,170 kilometres from Bruderheim, Alberta, to a new marine terminal to be built at Kitimat, British Columbia (BC) (Government of Canada, 2013). The project aims to transport approximately 525,000 barrels daily to the BC coast through one pipeline, while the second pipeline will transport liquid condensate from vessels to Alberta for oil sands production. However, the project faces two major environmental challenges. First, the pipelines would traverse a rough and untamed landscape. Moreover, Kitimat, the site of the proposed marine terminal, is located within a long, narrow fjord (Douglas Channel) characterized by an ecologically sensitive environment, sharp bends, and limitations related to weather and visibility ("Northern Gateway pipeline project," 2013; Swift et al., 2011).

The federal Joint Review Panel (National Energy Board and Canadian Environmental Assessment Agency) recommended the project's approval contingent upon 209 legally binding conditions, which the Harper cabinet subsequently accepted on June 17, 2014. Concerns about the NG's environmental risks to British Columbia's communities, rivers, and Pacific coastline, as well as its broader implications for Canada's greenhouse gas emissions, sparked unusually broad, multi-scalar opposition.

Despite the project's projected economic benefits and Enbridge's intensive propaganda efforts, the Kitimat residents rejected the project in a plebiscite held in April 2014 (Bowles & MacPhail, 2017; Gilchrist, 2014). At the national and provincial level, the Angus Reid Institute (2014) conducted a public opinion poll following the federal approval of the project, which revealed that the decision was deemed "wrong" by 34% of Canadians, and in British Columbia, the percentage increased to 40%. Moreover, Indigenous-law declarations, notably the Coastal First Nations Declaration and the Save the Fraser Declaration, were endorsed by more than 130 First Nations, who rejected pipelines and crude tankers through their territories (West Coast Environmental Law, 2012). These declarations constituted a jurisdictional challenge that was based on Indigenous authority rather than conventional stakeholder dissent (McCreary & Milligan, 2014).

Previous research (e.g., Dusyk et al., 2018; Gunster & Neubauer, 2019; Neubauer & Gunster, 2019) has identified a tension between national and local interests that helps explain the project's prolonged public controversy. Opponents focused on regionally grounded issues—salmon, coastal sovereignty, and Indigenous jurisdiction—contrasting with a federal/export-oriented "nation-building" narrative; mainstream media directed public attention more to local spill risks than to climate change; municipal/First Nations coalitions utilized local votes and legal tactics to undermine the project's legitimacy. The Government of British Columbia (2012), responding to ongoing public pressure, established five provincial conditions for the eventual approval of the project: a world-class spill response on land and sea, a completed environmental review, accommodation of Indigenous rights and benefits, and a "fair share" for British Columbia.

In September 2014, amid ongoing protests over NG, Enbridge launched the "Life Takes Energy" campaign as a bid to counter negative publicity (Cattaneo, 2014). The campaign sought to decouple pipelines from environmental risks and Indigenous sovereignty claims by shifting to lifestyle branding—presenting the significance of energy within intimate scenes of everyday life. It consisted of advertisements that ran on television, in print, and through social media. The multimedia content on the campaign's home page (http://www.enbridge.com/lifetakesenergy), which served as the basis for the creation of promotional messages in other communication channels, is the subject of the analysis that follows. Although as of 2025, the web page has been taken down by Enbridge, its non-video content can still be accessed via the Wayback Machine (https://web.archive.org/), a digital archive of the World Wide Web maintained by the Internet Archive, an American nonprofit organization. The following section also examines screenshots of the campaign's video advertisements, which the author downloaded in 2018 while they were accessible on the Enbridge website.

2.3 Life Takes Energy

The "Life Takes Energy" (LTE) campaign aimed to shift public attention away from politically charged narratives about NG and toward intimate, positive narratives about everyday life. It articulated Enbridge's corporate mission in a tripartite formula: a universal claim ("life requires energy"), a series of visuals illustrating appealing daily scenarios, and textual elements linking these scenarios to Enbridge's infrastructure and services (e.g., "when the energy you invest in life meets the energy we fuel it with, sweet things happen"). The formula legitimized "petro-prosperity" by associating Enbridge and its core services with the conditions of middle-class comfort (Huber, 2013).

Viewed through a discourse-historical lens (Reisigl, 2017; Reisigl & Wodak, 2016), LTE's major strategies were both textual and visual: the abstract, universal notion of "energy" replaced specific, contentious terms like oil, gas, and oil sands, while polished domestic scenes—baking, family warmth, daily commutes—were staged as proof that life depends on uninterrupted energy flows. Discursive choices, such as the second-person address ("you" invest energy in life) and the corporate "we" that "delivers" energy safely, encoded neoliberal lifestyles (e.g., automobility-enabled convenience and consumerism) as common sense rather than as historical, political outcomes. These strategies exemplified the lifestyle turn (McCurdy, 2018) by constructing a depoliticized naturalization of carbon dependency: energy consumption is given prominence, fossil fuel extraction is backgrounded, and pipelines are portrayed as morally neutral infrastructure facilitating ordinary happiness.

The core tactics of LTE's advertisements were as follows. First, they borrowed the visual grammar of food advertising and aspirational lifestyle branding to domesticate energy production and consumption (Cozen, 2010; Thomlison, 2019). Second, they framed energy as an enabling essence and presented Enbridge as a safe, reliable provider, while portraying customers as diligent, caring, and deserving. In contrast, contentious topics—such as Indigenous sovereignty and the environmental dangers and impacts of oil sands and pipelines—were absent. Third, the advertisements relied on two topoi: the topos of necessity (life requires energy; therefore, energy providers and pipelines are necessary) and the topos of consequence (if energy flows, good things happen: cupcakes, family time, warmth). They downplayed that Enbridge's energy product is based on fossil fuels transported through contested pipelines.

What Is "Oil Sands"?

LTE rendered oil sands discursively invisible by defining the commodity Enbridge delivers to customers as "energy"—an essence presented as value-neutral and inevitable. The campaign's lexical and visual choices obscured fossil fuels' troubling connotations (bitumen, pipelines, tankers, etc.) and anchored Enbridge's corporate identity in utopian, affect-rich scenes where the social and ecological relations of extraction are erased.

For example, the "Enbridge Story," which appeared as the lead feature video on the LTE homepage (Figs. 2.1 and 2.2), opens with the single word "energy," then moves through a rapid montage that mixes everyday scenes—a lit gas stove, a car driving through a suburb, someone barbecuing in a park, a couple strolling downtown, a mother and child reading by lamplight, etc.—with images of energy, or energy-powered infrastructure, such as an Enbridge-branded wind turbine, a control room, pipeline construction, a stacked freeway interchange, and a night-lit city skyline, before closing on the Enbridge logo with the tagline "life takes energy." Along with the montage, a friendly, warm male voice states that:

> Life takes energy. Energy to create the world we call home and fuel our quality of life. Energy to make life more comfortable and to help people to live their lives to the fullest. We dedicated ourselves to connecting people to the energy we all need. Now, as energy changes, we deliver our customers the energy from the Earth and harness the power of the sun and the wind. [...]

Although Enbridge operates the largest oil and gas pipeline networks in North America, the montage and accompanying voiceover associate energy with oil and gas only implicitly. Instead, energy is presented as the foundation of mobility, agriculture, middle-class life, and intimate care. The inserted images of energy or energy-powered infrastructure depict

Fig. 2.1 Screenshot of LTE Homepage (*Source* Internet archive, capture of website version 2024 January 18)

Fig. 2.2 Screenshots of the "Enbridge Story" (*Source* Internet archive)

Enbridge as a dependable custodian. By blending fossil fuel-coded scenes (pipelines and freeways) with renewables (wind turbines and solar panels) and avoiding contested extraction sites, the advertisement dissolves "oil sands" into the neutral abstraction "energy" and reattaches it to desirable lifestyles.

Fig. 2.3 Screenshots of "E = best dad ever" (*Source* Internet archive)

The LTE home page's second feature video, "E = best dad ever," further demonstrates how LTE positively linked Enbridge to the energy that people use in their homes (Fig. 2.3). The video stages an intimate, warmly lit scene in which a father turns on a gas knob to prepare dinner for his daughter. The father then turns on a dining room's light and begins sharing food with his daughter. The presentation of dining as a loving and caring moment highlights the home's sensory delights. Heat and light are decoupled from fossil fuel consumption and reframed as the affective infrastructure of love, competence, and safety. The video's accompanying text further emphasizes Enbridge's role as a benign, ever-present enabler of everyday family life and household energy use as an unproblematic condition for warmth and joy:

> We didn't plan a meal filled with foods that would make her smile. And we didn't make sure everything was absolutely perfect. But we did help you cook the meal that reminded her just how special she is. At Enbridge, we believe when the energy you invest in life meets the energy we fuel it with, beautiful things happen.

Who Are "We" in Relation to Oil Sands?

LTE frequency employed "we" to refer to Enbridge and its employees and emphasized Enbridge's intimate relationship with customers: your diligent care of family plus our safe, reliable energy delivery equals quality of life. This pairing linked everyday moments of care and convenience to corporate stewardship, positioning customers as neoliberal subjects deserving of comfort for their hard work (Huber, 2013) while branding Enbridge as a reliable provider of frictionless energy flows. Textually and visually, the construction of corporate stewardship was consistent throughout the campaign's promotional materials.

Enbridge's (2014) 2014 year-end blog post, for instance, explicitly collapses the corporate-family distance by stating that "when the energy you invest in life meets the energy we fuel it with, a holly, jolly holiday season happens" (para. 2). The post then reiterates the "life is fuelled by Enbridge" message by noting that "we bring you more than just energy" (Enbridge, 2014, para. 4); "we play a small but essential role in connecting North Americans to the quality of life they've come to expect" (para. 6); and "we also believe that it's important to remember, and consider, those beautiful life moments when your energy meets ours" (para. 11). Echoing the post's framing of Enbridge as a morally positive custodian and of the energy it delivers as the "energy of life," the header image presents a collage (a junior hockey team, popcorn, pastries, etc.) codifying the "E=" icon's link to ordinary moments (Fig. 2.4).

In short, the post enacts a double naturalization by framing Enbridge and its customers as sharing a cosy relationship: energy use appears constitutive of ordinary life, while the contingent question of how that energy is sourced is erased, rendering fossil inputs interchangeable with the neutral category "energy." Other LTE campaign messages also reiterated the double naturalization. Videos like "E = awestruck" and "E = making memories" encourage viewers to consider the energy that enables family vacations and school field trips (Fig. 2.5), thereby expanding the campaign's moral horizon to everyday convenience.

Fig. 2.4 Screenshot of Enbridge, 2014 (*Source* @Enbridge Blog)

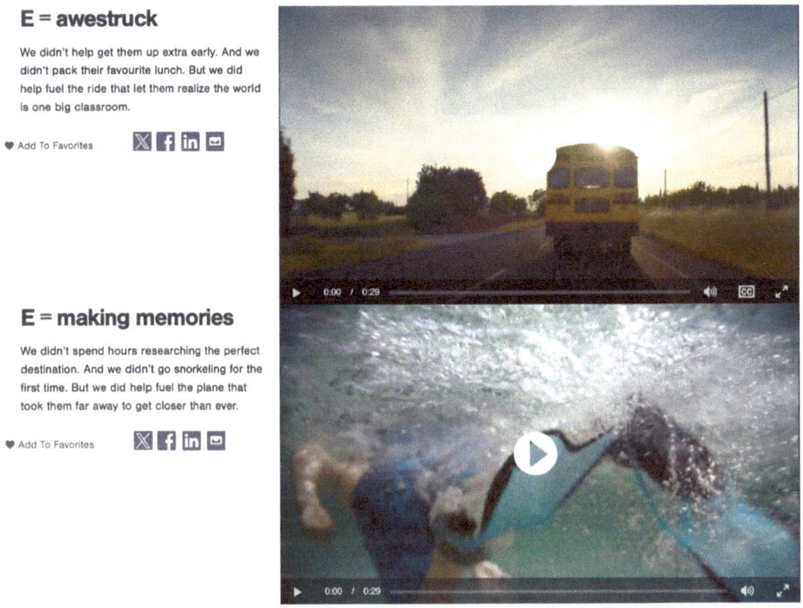

Fig. 2.5 Screenshots of "E = awestruck" and "E = making memories"

While videos such as "E = making memories" show joyful moments of daily life, the messages conveyed in their accompanying texts are nonetheless political (e.g., "We didn't spend hours researching the perfect destination. And we didn't go snorkelling for the first time. But we did help fuel the plane that took them far away to get closer than ever."). By discursively obscuring the boundary between Enbridge and family, LTE substituted consumer ethics (be diligent, be grateful for reliable service) for political economy (who decides energy futures, on what terms) and framed criticism of the corporation's environmental liabilities as a threat to daily pleasures. In short, the "we" in LTE established a responsible provider that caters to ordinary households, thereby normalizing fossil fuel dependency as common sense and excluding questions regarding carbon footprints and alternatives from discussions.

Who Are "They"?

Instead of explicitly naming antagonists, LTE constructed them as potential disruptors—cold, stillness, or inconvenience—that energy provided by Enbridge shields families from. Environmental concerns and critics of NG were absent from LTE's campaign materials: in videos like "E = best dad ever," familial warmth is lauded without reference to pipelines, oil sands, or any opposing viewpoints. Meanwhile, the repeated assertion that "we deliver the energy that fuels quality of life" shifted the focus of contention onto the threat of disrupted daily comforts. In light of the controversies surrounding NG, this neglect of antagonists was a calculated move to circumvent major opponents—Indigenous communities asserting sovereignty and coastal communities concerned about spill risk—and to downplay conflicts over climate, ecology, and rights.

What Are "We" Going to Do with Oil Sands?

The outlook set by LTE was the continuation of fossil fuel dependency with incremental improvements. LTE pledged to "fuel people's quality of life" safely and reliably, while also implementing incremental improvements in corporate social responsibility, stewardship, and safety. However, it did not address the structural transition toward renewables or the curtailment of oil and gas infrastructure expansion. Enbridge's, 2014 annual review presented this telos as:

"Enbridge exists to fuel people's quality of life. [...] By transporting, distributing, and generating energy, we play a critical role in enabling the economic well-being of North Americans. We deliver energy to where it's needed most—reliably, efficiently, and with the safety of our employees, the public, and the environment in mind" (Enbridge, 2015, p. 4).

To secure public consent for this outlook, Enbridge, in addition to utilizing the lifestyle branding mentioned above, positioned itself as a leader of energy transition. The 2014 annual review noted that the company "has interests in wind, solar, geothermal, and waste heat recovery facilities with a total generating capacity of more than 2,200 megawatts of emissions-free electricity" (Enbridge, 2015, p. 13). In 2014, it invested in projects in Keechi (Texas), Blackspring Ridge (Alberta), and Lac-Alfred/Massif-du-Sud (Quebec). However, these assets were presented as complementing a broader expansion of oil

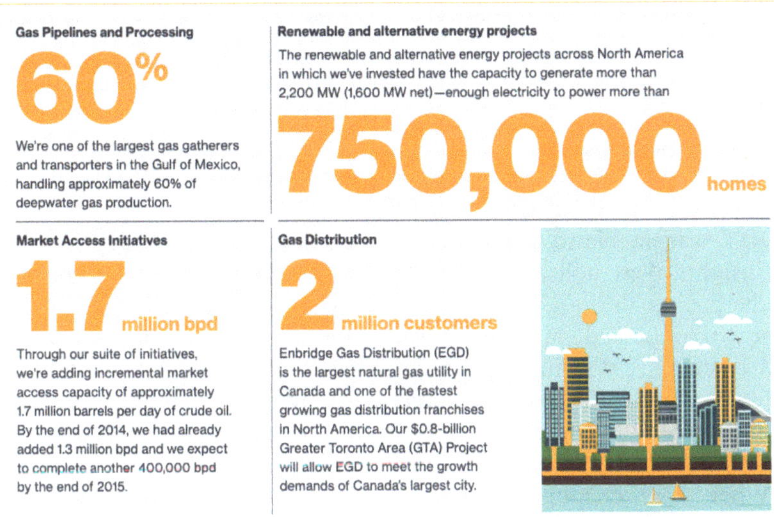

Fig. 2.6 Enbridge's core business, p. 10

and gas projects, not as a pathway to shift the company's core business (Fig. 2.6). Thus, renewables functioned rhetorically as additive co-investments within an outlook premised on expanding and optimizing the company's primary business of delivering gas and oil.

2.4 Assessment: Carbon Dependency and Everyday Life

All things considered, LTE portrayed fossil fuels as the lifeblood of consumer culture: warmth in kitchens, mobility on roads, coziness in living rooms, and community in seasonal rituals. The campaign exemplified the "selling oil sands without oil sands" approach (McCurdy, 2018) by subtly promoting Enbridge's core business—oil and gas transportation—through the neutral concept of "energy," legitimizing dependence on fossil fuels while erasing the social relations and negative impacts of extraction.

Viewed through the lenses of petroculture and political economy, LTE foregrounds privatized satisfactions—home comfort, automobility, and convenience—as the central concerns in public discourse about society's energy futures, while defining oil and gas as the foundation of daily life. LTE thus followed the consent track of petro-hegemony (LeQuesne, 2019): by aligning prosperity with infrastructural expansion, it presented NG and similar projects as custodians of life rather than contested choices among possible energy paths. Pipeline construction was framed as a continuity of ordinary happiness, narrowing the imaginative space for collective decision-making about energy transition.

LTE was unable to change the public's perception of NG. According to Insights West polls, British Columbians held a net-negative opinion of the project during 2015–2016 (Canseco, 2014; Crawford, 2016). In June 2016, the Federal Court of Appeal quashed the federal approval NG had received, citing inadequate consultation with Indigenous communities (de Souza, 2016). This underscored that legal and constitutional claims—especially Indigenous jurisdiction—could not be neutralized by lifestyle advertising.

When LTE encountered politicized audiences, its limitation as an apolitical approach was also revealed. Following a petition signed by 28,000 people, Tim Hortons, a Canadian coffee chain, removed LTE ads from its in-store TV screens in June 2015 (Armstrong, 2015). However, supporters of NG then launched a counter-boycott in response to this decision. Thus, LTE was perceived as pipeline advocacy by both NG supporters and opponents, which reduced the effectiveness of its use of lifestyle branding and positive sentiments. In other words, by not naming "they," LTE sought to gain NG's social license by portraying opposition as "anti-ordinary." Despite this, the public continued to perceive the campaign as interfering in ongoing disputes over Indigenous rights, the environment, ecology, and climate change (Dusyk et al., 2018; Gunster & Neubauer, 2019; Neubauer & Gunster, 2019).

This said, one message constructed by LTE nevertheless endured, even after the cancellation of NG. By universalizing "energy" as the condition of everyday life, LTE helped entrench a rhetorical pivot that accuses oil sands critics of being "hypocrites" who enjoy energy's benefits while opposing its infrastructure (Gunster et al., 2018). This message later fused with populist sentiment and evolved into a powerful discursive framework defending oil sands expansion, which I will discuss in more detail in the upcoming chapters.

References

Angus Reid Institute. (2014, June 18). *Northern Gateway: Canadians split on federal decision; but majority say pipeline will be built*. https://angusreid.org/northern-gateway-canadians-split-on-federal-decision-but-majority-say-pipeline-will-be-built

Armstrong, J. (2015, June 4). Tim Hortons faces backlash after yanking Enbridge ads from stores. *Global News*. https://globalnews.ca/news/2036702/tim-hortons-faces-backlash-after-yanking-enbridge-ads-from-stores

Barney, D. (2017). Who we are and what we do: Canada as a pipeline nation. In S. Wilson, A. Carlson, & I. Szeman (Eds.), *Petrocultures: Oil, politics, culture* (pp. 78–119). McGill-Queen's University Press.

Bowles, P., & MacPhail, F. (2017). The town that said "No" to the Enbridge Northern Gateway pipeline: The Kitimat plebiscite of 2014. *The Extractive Industries and Society, 4*(1), 15–23. https://doi.org/10.1016/j.exis.2016.11.009

Canseco, M. (2014, August 7). Most British Columbians oppose the Enbridge Northern Gateway pipeline. *Vancouver Observer*. https://www.vancouverobserver.com/news/most-british-columbians-oppose-enbridge-northern-gateway-pipeline.html

Cattaneo, C. (2014, September 22). Enbridge Inc aims to stem negative publicity with Life Takes Energy rebranding campaign. *Financial Post*. https://financialpost.com/commodities/energy/enbridge-inc-aims-to-stem-negative-publicity-with-life-takes-energy-rebrandin-campaign

Cozen, B. (2010). This pear is a rhetorical tool: Food imagery in energy company advertising. *Environmental Communication, 4*(3), 355–370. https://doi.org/10.1080/17524032.2010.499212

Crawford, E. (2016, March 2). Pipeline popularity dropping in B.C.: Insights West. *Business in Vancouver*. https://www.biv.com/news/resources-agriculture/pipeline-popularity-dropping-bc-insights-west-8245660

de Souza, M. (2016, June 30). Court quashes Enbridge pipeline, slams Harper government for ignoring First Nations. *National Observer*. https://www.nationalobserver.com/2016/06/30/news/court-rejects-enbridge-pipeline-approval-slams-harper-government-ignoring-first

Dusyk, N., Axsen, J., & Dullemond, K. (2018). Who cares about climate change? The mass media and socio-political acceptance of Canada's oil sands and northern gateway pipeline. *Energy Research & Social Science, 37*, 12–21. https://doi.org/10.1016/j.erss.2017.07.005

Enbridge. (2014, December 23). *Home for the holidays: Celebrating a holly, jolly Christmas*. https://www.enbridge.com/stories/archived/2014/december/life-takes-energy-home-for-the-holidays

Enbridge. (2015). *Enbridge Inc. 2014 annual review*. https://www.enbridge.com/~/media/Enb/Documents/Investor%20Relations/2015/2014_ENB%20AnnualReview.pdf

Gilchrist, E. (2014, April 13). Kitimat says 'no' to Northern Gateway. *The Tyee*. https://thetyee.ca/News/2014/04/13/Kitimat-Vote/

Government of Canada. (2013). *Connections: Report of the joint review panel for the Enbridge Northern Gateway project*. https://publications.gc.ca/site/eng/456575/publication.html

Government of British Columbia. (2012, July 23). *British Columbia outlines requirements for heavy oil pipeline consideration*. https://news.gov.bc.ca/releases/2012ENV0047-001074

Gunster, S., Fleet, D., Paterson, M., & Saurette, P. (2018). "Why don't you act like you believe it?": Competing visions of climate hypocrisy. *Frontiers in Communication, 3*, Article fcomm.2018.00049. https://doi.org/10.3389/fcomm.2018.00049

Gunster, S., & Neubauer, R. J. (2019). (De)legitimating extractivism: The shifting politics of social licence. *Environmental Politics, 28*(4), 707–726. https://doi.org/10.1080/09644016.2018.1507290

Hjarvard, S. (2008) The mediatization of society. A theory of the media as agents of social and cultural change. *Nordicom Review 29*(2): 105–134. https://doi.org/10.1515/nor-2017-0181

Huber, M. T. (2013). *Lifeblood: Oil, freedom, and the forces of capital*. University of Minnesota Press.

LeQuesne, T. (2019). Petro-hegemony and the matrix of resistance: What can Standing Rock's water protectors teach us about organizing for climate justice in the United States? *Environmental Sociology, 5*(2), 188–206. https://doi.org/10.1080/23251042.2018.1541953

MacDonald, G. (2017). Containing oil: The pipeline in petroculture. In S. Wilson, A. Carlson, & I. Szeman (Eds.), *Petrocultures: Oil, politics, culture* (pp. 36–77). McGill-Queen's University Press.

McCreary, T. A., & Milligan, R. A. (2014). Pipelines, permits, and protests: Carrier Sekani encounters with the Enbridge Northern Gateway Project. *Cultural Geographies, 21*(1), 115–129. https://doi.org/10.1177/1474474013482807

McCurdy, P. (2018). From the natural to the manmade environment: The shifting advertising practices of Canada's oil sands industry. *Canadian Journal of Communication, 43*(1), 33–52. https://doi.org/10.22230/cjc.2018v43n1a3315

McCurdy, P. (2023). The rise and fall of the Synthetic: The mediatization of Canada's oil sands. *International Journal of Cultural Studies, 26*(4), 427–444. https://doi.org/10.1177/13678779231159697

Neubauer, R., & Gunster, S. (2019). Enemies at the Gateway: Regional populist discourse and the fight against oil pipelines on Canada's West Coast. *Frontiers in Communication, 4*, article fcomm.2019.00061. https://doi.org/10.3389/fcomm.2019.00061

"Northern Gateway pipeline project: 6 things to know." (2013, December 19). *CBC News*. https://www.cbc.ca/news/canada/northern-gateway-pipeline-project-6-things-to-know-1.2469561

Paterson, M. (2007). *Automobile politics: Ecology and cultural political economy*. Cambridge University Press.

Reisigl, M. (2017). The discourse-historical approach. In J. Flowerdew, J. E. Richardson (Eds.), *The Routledge handbook of critical discourse studies* (pp. 44–59). Routledge. https://www.routledgehandbooks.com/doi/10.4324/9781315739342.ch3

Reisigl, M., & Wodak, R. (2016). The discourse-historical approach (DHA). In R. Wodak & M. Meyer (Eds.), *Methods of critical discourse studies* (3rd ed., pp. 23–61). Sage.

Simpson, S. (2010, November 30). Yogurt comparison no joke to oil sands opponent. *Vancouver Sun*. https://vancouversun.com/news/staff-blogs/yogurt-comparison-no-joke-to-oil-sands-opponent

Swift, A., Lemphers, N., Casey-Lefkowitz, S., Terhune, K., Droitsch, D. (2011, November). *Pipeline and tank trouble: The impact to British Columbia's communities, rivers, and Pacific coastline from tar sands oil transport*. Pembina Institute. https://www.pembina.org/reports/nrdc-pipeline-and-tanker-trouble-for-web.pdf

Szeman, I. (2017). Conjectures on world energy literature: Or, what is petroculture? *Journal of Postcolonial Writing, 53*(3), 277–288. https://doi.org/10.1080/17449855.2017.1337672

Takach, G. (2013). Selling nature in a resource-based economy: Romantic/extractive gazes and Alberta's bituminous sands. *Environmental Communication, 7*(2), 211–230. https://doi.org/10.1080/17524032.2013.778208

Thomlison, A. (2019). From pipeline to plate: The domestication of oil sands through visual food analogies. In H. Greaves & D. Beard (Eds.), *The rhetoric of oil in the twenty-first century: Government, corporate, and activist discourses* (pp. 189–210). Routledge.

West Coast Environmental Law. (2012). *First Nations that have declared opposition to proposed Enbridge tanker & pipeline project*. https://www.wcel.org/publication/first-nations-have-declared-opposition-proposed-enbridge-tanker-pipeline-project

CHAPTER 3

"Keep Canada Working": Promoting Resource Nationalism via Symbolic Nationalization

Abstract Focusing on Alberta's 2018 "Keep Canada Working" campaign during the Trans Mountain Pipeline Expansion controversy, the chapter identifies a rhetoric of "symbolic nationalization" that recoded a foreign-capital-dominated private sector as a national commons. This resource-nationalist imaginary mobilized unity and sovereignty tropes to mask uneven risks and benefits, promising prosperity while obscuring material contradictions on the ground. The chapter situates these strategies within petro-state dynamics and the "resource nationalism" literature to show how national belonging is enlisted to normalize extractivism.

Keywords Persuasion · Petro-state · Resource nationalism · Social license · Symbolic nationalization · Trans Mountain

This chapter explores the controversy surrounding the Trans Mountain Pipeline expansion (TMX) project between 2015 and 2018. During this period, Alberta was governed by the provincial New Democratic Party (NDP), which, compared to the provincial Progressive Conservative Party that governed the province for over four decades (1971–2015), took a more progressive approach to defending the interests of the oil sands industry. The chapter offers a critical analysis of "Keep Canada Working,"

S. Chen, *Petro-Rhetoric in Canada*, Palgrave Studies in Media and Environmental Communication,
https://doi.org/10.1007/978-3-032-12964-2_3

a short-lived campaign launched by the Alberta NDP government in 2018 with the aim of amassing broader nationwide backing for TMX. The analysis elucidates the rhetoric of "symbolic nationalization," which, by concealing the fact that the oil sands industry is a private sector heavily controlled by foreign capital, seeks to define extractivism as part of the Canadian identity. Overall, the chapter underscores the stark contradictions between the promises of homogenizing populist appeals and the actual reality on the ground.

3.1 Resource Nationalism and Its Rhetorical Appeals

The scholarly literature concerning energy governance, policy, and politics has extensively explored the relationship between fossil fuel extraction and the state, analysing topics such as the emergence of the oil assemblage and the political influence exerted by fossil fuel corporations (Mitchell, 2011; LeQuesne, 2019; Taft, 2017). A key insight provided by this body of literature is that the establishment of state-corporate nexuses in economies reliant on natural resources often results in a phenomenon known as the "resource curse" (Karl, 1997; Kennedy, 2014), which depicts how governments structure their administrative systems, economy, politics, and policy priorities to cater to the requirements of extractive industry development. Karl's (1997) original petro-state thesis posits that dependence on oil revenues transforms fiscal and political institutions into a rentier order. Hydrocarbon revenues dominate public finance and foreign exchange, power consolidates in the executive, discretionary patronage and budget constraints increase, the tax-representation bargain deteriorates, and price volatility leads to pro-cyclical spending, indebtedness, and Dutch-disease pressures. These factors collectively reinforce coalitions and policies prioritizing the development of extractive industries.

According to Mitchell (2011), there is frequently a convergence of interests between extractive industries and states because, in countries endowed with abundant resources, the legitimacy of governments and their capacity to fulfil the expectations of citizens are contingent on resource-driven economic growth. Meanwhile, state support is crucial for extractive industries due to its provision of bureaucratic structure and state-sanctioned violence, which ensure secure extraction and transportation of natural resources.

Mitchell (2011) further argues that carbon energy systems are not politically neutral infrastructures but "machines" that organize power: the material properties of coal and oil and the networks that move them shape what kinds of politics are possible. For example, as oil became the primary energy source for capitalism, it reorganized energy production and transportation into capital-intensive, long-distance pipelines, shipping lanes, and offshore fields that dispersed workers and narrowed "chokepoints" potentially targeted by labour movements, enabling firms and states to bypass labour pressures.

Meanwhile, governments rely on a continuous supply of cheap energy to bolster their legitimacy. They achieve this by expanding expert bureaucracies in finance, planning, and other areas; subsidizing consumption; and deploying police and military forces to protect critical energy supply chains. In short, the convergence of interests between states and extractive industries is not merely ideological but engineered by socio-technical design.

As a result of the alliance between state and corporate actors, the boundary between public and private interests becomes ambiguous during the expansion of resource extraction frontiers. While this ambiguity is highlighted by concepts like "extractivism" and "petro-state" in scholarly literature (e.g., Chagnon et al., 2022; Kennedy, 2014), it is often downplayed in public discourse. In business news and corporate press releases, for instance, the term "resource nationalism" is preferred over "extractivism" and "petro-state" when referring to state interventions pertaining to the extraction, processing, and sale of natural resources. For example, in the following excerpt from a commentary published by Reuters, Indonesia's tightening of export permits and Myanmar's mine suspensions were framed as supply risks driven by "resource nationalism":

> Even assuming a speedy resumption of Indonesian exports and mining in Myanmar, tin supply over the next few months looks challenging. The longer-term threat is of future supply disruption as resource nationalism drives both governments further down the road of export controls. (Home, 2024, para. 27–28)

Resource nationalism has gained prominence in both scholarly and public discussions since the 2000s. This is mainly due to the prolonged boom in primary commodity prices from 2003 to 2011, which prompted numerous efforts to nationalize resource sectors in Global South countries

(Pryke, 2017; Veltmeyer & Bowles, 2014). By "primary commodities boom," I mean a sustained, broad-based upswing in the real prices of fuels, metals, and agricultural products—interrupted only by the 2008–09 global financial crisis and often described as a commodity "super-cycle" because of its duration and cross-sector scope (Radetzki, 2006). The demand-side drivers of this boom were rapid urbanization and infrastructure investment in China and other emerging economies, while supply constraints, long investment lags, and policy shifts (e.g., biofuels) tightened global markets.

In writings on resource policy, mentions of resource nationalism reflect a dichotomy between liberalism and statism: they revolve around a state's relative control over and benefit from natural resources as opposed to private corporations (Pryke, 2017). The dichotomy portrays national governments as intermediaries in global markets, employing discretionary and selective policies in order to secure political and/or economic advantages that would not be possible if markets were left unregulated (Wilson, 2015).

The liberalism–statism dichotomy corresponds to a range of mechanisms—ownership regulations, fiscal conditions, and trade and investment restrictions—by which states endeavour to reallocate rents and authority from private or foreign entities to public actors (Arbatli, 2018; Pryke, 2017; Wilson, 2015). The statist approach generally includes elevated royalties and windfall taxes, obligatory state equity, export restrictions, and more stringent scrutiny of foreign investment. In contrast, the liberal approach seeks to reverse these measures through privatization, tax incentives, deregulation, and other policies aimed at positioning governments as "market referees" rather than market participants. The various policy alternatives within the dichotomy are well exemplified by Latin America in the 2000s. Buoyed by the rise in primary commodity prices, a number of Latin American governments (e.g., Bolivia, Venezuela, etc.) modified their natural resource policies. Rather than seeking complete expropriation, they implemented hybrid arrangements to reconcile state and private interests (Cypher, 2010).

However, resource nationalism becomes an ill-defined term when the shifting political geography of extractive industries is taken seriously, specifically considering the fact that new frontiers of extraction are emerging in increasingly extreme spaces and have sparked intensified political struggles (Arbatli, 2018; Childs, 2016; Koch & Perreault, 2019; Pryke, 2017). For example, the planned rare-earth/uranium mining at

Kvanefjeld in Greenland has caused division among the local population and sparked opposition based on worries about radiological risk and environmental harm (Henriques & Böhm, 2022). The complexity of the Kvanefjeld case arises from the intersection of sub-national autonomy, uranium bans, and community vetoes with foreign investment and intensifying geopolitical tensions concerning "critical minerals," resulting in a political landscape that is simultaneously local, national, and geopolitical rather than straightforwardly "nationalist."

In reality, the discourse of resource nationalism constitutes multiple and often contradictory narratives, underlying which are very different motives and policies by governments. The limitations of business news and corporate press releases in capturing this complexity result in a tendency to depoliticize the normative aspects of their claims regarding concepts like "nationalism," "territory," and "sovereignty," thereby neglecting the political process that underpins the production of these concepts.

For example, in business media coverage on the nationalization of resource industries by governments of developing countries (e.g., Home, 2024), the term "resource nationalism" is typically used pejoratively to describe how nationalism jeopardizes the supply of vital resources to industrialized countries and multinational corporations. Such manifestations of prejudice towards political orientation have been problematized by critical geographers (e.g., Childs, 2016; Haslam & Heidrich, 2016), who propose that resource nationalism should be used in a neutral manner to denote an increasingly diverse set of development policies and practices (e.g., hybrid or post-neoliberal forms of governance) that are inspired by economic nationalism. States embracing "resource nationalism" do not solely oppose liberalization; instead, they often adjust ownership, taxation, and control while maintaining market principles, resulting in diverse "post-/neo-extractivist" configurations.

In line with the theoretical insights offered by critical geography and contiguous fields, this chapter defines resource nationalism as a political discourse mobilized by diverse actors and producing uneven socio-environmental outcomes. It emphasizes that "the people of a given country, rather than private corporations or foreign entities, should benefit from the resources of a territorially-defined state" (Koch & Perreault, 2019, p. 612). This definition attends to not only the distribution of benefits derived from resource extraction but also the expression of collective belonging through sacralizing natural resources. From a discursive perspective, resource nationalism operates as a repertoire of narratives that, through media practices, branding, and affective claims about belonging and security, construct natural resources like fossil fuels to appear as the people's commons rather than corporate assets (Szeman, 2013; Wilson et al., 2017). Regarding oil, resource nationalism bolsters (1) lifestyle advertisements' assertion that it is associated with freedom, mobility, and middle-class security (Huber, 2013), as well as populist antagonisms (e.g., "our oil" and "foreign-funded environmental lobby") that create a dichotomy of energy citizens versus external/internal enemies (Gunster & Saurette, 2014).

The incorporation of a discursive perspective departs from the economic determinism that is commonly associated with analyses of boom-and-bust cycles in energy policy. It enables more scholarly attention to (1) when, where, why, and to what effect resource-related public narratives function as a force shaping politics and policymaking, and (2) the significance of non-state and sub-national actors in generating such narratives, thereby influencing public perceptions of the construction of an imagined community.

Resource nationalism rhetoric is closely linked to fossil fuel extraction and the denial of its adverse consequences (Conversi, 2020; Koch & Perreault, 2019). In both developing and developed countries, such rhetoric has been deployed by government actors and their allies to gain public support for developing energy infrastructure or to seize control of fossil fuel reserves. There has been a discernible trend in recent years where the celebration of resource nationalism merges with populism. Within Europe, for instance, the intertwining of nationalist sentiments with energy politics has been highlighted by public debates surrounding the coal and shale gas industries in Poland (Lis, 2018; Materka, 2011; Żuk et al., 2024). Resource nationalism has fused with right-populist imaginaries, collectively positioning the coal and (anticipated) shale gas industries as pillars of sovereignty, security, and social stability. Jaspal et al. (2014) conducted an analysis of the Polish press' coverage of fracking and discovered that energy security and independence were key drivers of the Polish media's favourable and enthusiastic perspectives on fracking. The Polish media normalized a pro-fracking stance by foregrounding energy independence from Russia and the promise of national revitalization. In contrast, criticisms and concerns about the environmental effects of fracking were characterized as "anti-Polish," "foreign-influenced," and counter-normative, leading to the suppression of these perspectives (Jaspal et al., 2014; Żuk et al., 2024). For instance, EU climate governance and Polish environmental NGOs were consistently portrayed as antagonistic elites jeopardizing the welfare of "real Poles" in pro-government media narratives.

At the heart of resource nationalism's rhetorical appeals is the conceptualization of the nation-state as a cohesive, congruent, and homogeneous whole (Conversi, 2020). This allows a regime to pursue an aggressive agenda of controlling territorial resources while ignoring whether the lives of those living near the frontlines of extractive expansion are impacted. Put differently, nationalism's boundary-making and homogenizing imperatives prioritize energy "security" and rent distribution over ecological and environmental considerations, externalizing social-environmental costs to marginalized communities and undermining the cosmopolitan coordination necessary for effective climate mitigation. Consequently, there tend to be stark differences between what is promised by homogenizing nationalist rhetoric and the reality on the ground, with communities most severely affected socially and environmentally being the least to benefit from extractivism.

One case aptly illustrating such differences is Ecuador during the presidency of Rafael Correa (2010–2017). As part of the "Pink Tide" in Latin America, the Correa government rejected neoliberalism and launched a series of reforms aiming to improve Ecuador's economic equality and living standards, primarily financed through the nationalization of extractive industries (Riofrancos, 2020). The Correa government significantly increased its revenue from hydrocarbons by instituting state-controlled service contracts in the oil and gas sector and raising the windfall-profits tax on oil revenues, subsequently utilizing these rents to enhance public investment and social expenditure.

While it was unsurprising that Correa's reforms encountered opposition from corporations and domestic elites, Ecuador's resource nationalism also faced fierce resistance from many grassroots activists who were part of the coalition that supported Correa's leftist election platform. In Ecuador's so-called Copper Belt provinces of Morona Santiago and Zamora Chinchipe, local Indigenous communities mobilized against China-backed mega-mining projects, prompting a state of emergency and mass displacement in December 2016 (Quiliconi & Vasco, 2021; Riofrancos, 2020). Despite allegations of inadequate consultation and forced evictions, the mining project proceeded, demonstrating how national extractive projects shift risks to peripheral communities.

According to Riofrancos' (2020) analysis, the activists' firm anti-extractivism standpoint was driven by the massive social and environmental damage caused by extractive projects. To many communities threatened by intensifying extractive activities, the Correa government's self-proclaimed socialist agenda was little different from the colonial exploitation they had endured for centuries. These communities thus became a major political force pushing for the recognition of their Indigenous knowledge on land and waterways, new forms of democratic participation, and a post-neoliberal state with genuine commitments to decolonization and sustainability.

Resource nationalism and its accompanying rhetoric pose substantial threats to global endeavours aimed at mitigating climate change. Climate change, being an uncontainable phenomenon that transcends geographical and ethnic boundaries, requires states to work together in a manner that urgently demands the reduction of emotional tensions towards the Other (Conversi, 2020). This is at odds with state-sanctioned fossil fuel development, which, in the name of public interest, often colludes with transnational carbon capital and causes extensive damage to a nation's neighbours or to the rest of the world. Consider the case of Ecuador. The Yasuní-ITT initiative, which aimed to leave oil in the ground in exchange for international compensation, was abandoned by the Correa government in 2013. Instead, drilling was permitted in parts of the UNESCO biosphere reserve known as Yasuní National Park (Laastad, 2024; Riofrancos, 2020). This decision has a significant environmental impact, particularly greenhouse gas emissions and the threat to globally significant biodiversity, but it was justified by nationalist claims about funding social development.

Thus, critical criminologists such as White and Kramer (2015) argue that resource nationalism, when utilized to legitimize hydrocarbon extraction, should be considered a form of international aggression whose consequences cross borders and generations rather than a purely domestic matter. They advocate recognizing ecocide—environmental destruction committed by the state-corporate nexus through law and policy—and pursuing internationalized accountability commensurate with the transboundary nature of climate change.

Resource nationalism in Canada exhibits both similarities and distinctions in comparison to countries like Poland and Ecuador. Both Canada's political economy and the formation of its national identity have long been associated with the production and export of staples (Stanford,

2020). The staples thesis, which builds on the research findings of Harold Innis (1970) and Mel Watkins (1963), states that growth in an economy reliant on resources is structured around a series of "staples" whose backward, forward, and final-demand linkages influence institutions, diversification, and regional settlement. Staples can trap economies in primary export dependence and create a "staples state" focused on extraction, transportation corridors, and metropolitan-hinterland hierarchies.

The 2000s primary commodities boom led to unprecedented growth in resource extraction and export within Alberta's oil sands sector, characterized by significant capital intensity, extensive energy infrastructure, and influential state-corporate alliances that directed macroeconomic policy and public discourse towards export expansion (Kraushaar-Friesen & Busch, 2020; Stanford, 2020; Veltmeyer & Bowles, 2014). Akin to numerous countries in Latin America, the emergence of oil sands as a "super staple" has culminated in a multitude of negative consequences. These include, but are not limited to, detrimental impacts on the environment, spillover effects on other export sectors, excessive political sway exerted by fossil fuel industries, and violations of Indigenous sovereignty and rights (Gunton, 2020). Other classic effects of the "staples trap" also followed: regional inequality between producing and non-producing provinces, boom-bust fiscal cycles, infrastructure and regulatory lock-in, and crowding-out pressures on manufacturing due to currency appreciation. In short, the oil sands boom has confirmed the staple theory's contemporary relevance.

Unlike Latin America, where fossil fuels are framed as public goods because their production is dominated by nationalized corporations, Canada's petro-nationalism is unique in that ownership and control are diffused among private firms within a constitutional framework that delegated resource jurisdiction to provinces and did not include a federal oil company following Petro-Canada's 1991 privatization (Yusufali & Pratt, 2009). Alberta's oil sands industry is notable for disproportionately serving the interests of private corporations and investors due to the low royalties, taxes, and fees imposed on it (Taft, 2017). In Europe, a significant aspiration for energy security and autonomy is pivotal in propelling the expansion of unconventional fossil fuel extraction (Jaspal et al., 2014; Żuk et al., 2024), while Canada's energy sector predominantly emphasizes exports. The emphasis on market access and corridor development influences pertinent public narratives.

Overall, the effectiveness of Canadian petro-nationalism depends on how successful the petro-bloc, consisting of fossil fuel corporations and their allies, is at minimizing the significant environmental impact of oil sands, defending the substantial subsidies granted to the oil sands industry, and thwarting initiatives aimed at decreasing Canada's reliance on fossil fuel exports (Kraushaar-Friesen & Busch, 2020). A major discursive framework developed by the petro-bloc to accomplish these objectives is "symbolic nationalization," which, as I will explicate in the next section, "promotes extractivism as a national public good and positions critics as anti-Canadian and foreign to the body politic" (Gunster et al., 2021, p. 58).

3.2 The Trans Mountain Pipeline Controversy, 2015–2018

The original Trans Mountain pipeline began to operate in 1953 for transporting oil from Alberta to the West Coast. In 2012, its owner, the Canadian division of U.S. pipeline giant Kinder Morgan Inc., proposed the Trans Mountain Pipeline Expansion (TMX) project, aiming to twin the existing line and triple its capacity from approximately 300,000 barrels per day to 890,000 barrels per day (Canadian Press, 2024; Trans Mountain Corporation, n.d.-a). Besides upgrading and expanding the pipeline system, the project proposed to construct three new berths at the Westridge Marine Terminal in Burnaby, a city in the Lower Mainland region of British Columbia (BC). The expanded terminal was expected to handle about 34 tankers per month.

Similar to the Northern Gateway Pipelines (NG) project discussed in Chapter 2, TMX aimed to enable Alberta's oil sands industry to access "tidewater" and increase exports to Asian markets. Both projects encountered persistent opposition from environmental groups, Indigenous communities, municipalities, and concerned citizens, resulting in protracted litigation over consultation and environmental risks. In the case of TMX, increased oil sands production resulting from it was estimated to generate an additional 13–15 megatons of greenhouse gas emissions per year (Coast Protectors, n.d.). The pipeline upgrade and expansion would cross multiple areas with sensitive environments and species at risk. Additionally, although TMX was designed to terminate in the urbanized Burrard Inlet, the substantial increase in tanker traffic associated with it would result in significant adverse effects on southern resident killer whales and on the traditional Indigenous use associated with the species (Raincoast, 2019). In the event of a major oil spill, the ecology and environment of the Burrard Inlet and the Fraser River Estuary would be severely affected (Short, 2015). This concern was reinforced by the fact that the original Trans Mountain pipeline had already experienced numerous oil spills, notably the Westridge 2007 spill (Trans Mountain Corporation, n.d.-b), which required years of cleanup.

The significant environmental impacts of TMX have made it one of the most contentious issues in Canadian politics. The collective opposition it faced in British Columbia has been particularly fierce (Hoberg, 2016). In 2016, the BC government formally opposed the project, citing deficiencies in the proposal's emergency response preparedness for oil spills ("Trans Mountain pipeline expansion," 2016). The Lower Mainland's municipal governments took the same stance. For example, Derek Corrigan, the mayor of Burnaby at the time, openly declared that he was ready to end his political career by being arrested for protesting against TMX (Moreau, 2015). Among Indigenous communities, the Tsleil-Waututh First Nation, whose traditional territory spans the Burrard Inlet, played the most active role opposing the project: they were at the forefront of anti-TMX protests and conducted an independent assessment of the project, ultimately rejecting it in May 2015 based on their own laws (Stewart et al., 2023). Along with Indigenous communities, environmental organizations and citizen groups played important roles in mobilizing concerned citizens to join the anti-TMX movement. For example, the Dogwood Initiative, a BC-based grassroots organization, launched the "No Tankers" campaign, designed to keep oil tankers away from the BC coast. Over the campaign's history, its petition was "signed by 290,111 British Columbians and about 90,000 other Canadians" (Hoberg, 2016, p. 10).

As a result of the prolonged political division, TMX's review and approval process experienced several setbacks (Canadian Press, 2024). The National Energy Board (NEB) recommended approval of TMX in May 2016, with 157 conditions designed to mitigate the project's environmental risks (Canada Energy Regulator, 2024; Canadian Press, 2024). The federal cabinet led by Justin Trudeau then approved TMX on November 29, 2016, while officially rejecting NG, which was deemed not to be in the national interest.

Nevertheless, despite the federal approval, British Columbian opposition remained strong. On January 30, 2018, the BC government initiated a restriction on any increase in diluted bitumen shipments until further spill response studies are conducted, thereby intensifying uncertainty for TMX. Then, between March and April, protests escalated near the Westridge Marine Terminal, with at least 176 protesters arrested for violating a court injunction prohibiting demonstrations within five metres of the terminal (Anderson, 2018).

In light of persistent local dissent and escalating expenses resulting from project delays and uncertainty, Kinder Morgan Canada declared the suspension of non-essential expenditures on TMX on April 8 and set a deadline of May 31 to finalize agreements with stakeholders. On May 29, the federal government announced it would purchase the pipeline and expansion project from Kinder Morgan Canada for $4.5 billion, converting TMX into a publicly owned project. The acquisition marked a pivotal moment in the TMX controversy, providing proponents of TMX with the opportunity to redirect public discourse from "Kinder Morgan (a U.S. corporation) versus BC citizens" to "TMX as a matter of national interest." In this context, the Alberta government, led by the provincial New Democratic Party (NDP), launched the "Keep Canada Working" campaign.

3.3 Keep Canada Working

The "Keep Canada Working" (KCW) campaign aimed to make an impact in a socio-political context where TMX's future remained uncertain. Although the project had just been purchased by Ottawa, it still faced Indigenous legal challenges and opposition from municipal and provincial stakeholders in British Columbia, who continued to contest spill preparedness and cumulative environmental impacts. KCW opted to address these criticisms by emphasizing the public benefits associated with TMX, such as enhanced market access, fiscal returns, and interprovincial solidarity, thereby symbolically nationalizing the project (Gunster et al., 2021; Koch & Perreault, 2019).

Between May 2015 and April 2019, Alberta was under the governance of the provincial NDP, whose stance on climate change and carbon taxation had a significant impact on KCW's overall strategy. The Alberta NDP, a social democratic party situated on the centre-left of Canada's political spectrum, embraced a more progressive stance on climate change than its predecessor, the Progressive Conservatives, who were in power for over 40 years before 2015. The Alberta NDP pursued a distinct "climate policy-for-market-access" bargain with the federal government. Its 2015 Climate Leadership Plan (International Energy Agency, 2024) comprised four principal policy pillars: establish a pricing system on greenhouse gas emissions; by 2030, eliminate coal-generated electricity and derive 30% of electricity from renewable sources; impose a cap on oil sands emissions at 100 megatons annually; and decrease methane emissions from upstream oil and gas production by 45% by 2025, relative to 2014 levels.

The Alberta NDP presented these policies as the credible environmental measures required to gain national support for TMX, whose royalties and job creation are vital to the provincial economy. Furthermore, between 2017 and 2018, when Alberta and British Columbia were embroiled in disputes regarding TMX, the latter was also governed by its provincial NDP. This further constrained the Alberta NDP's ability to utilize adversarial narratives against its BC counterpart.

Accordingly, KCW followed the "information deficit" model (Sturgis & Allum, 2004): it framed the lack of public support for TMX as stemming from knowledge gaps (about pipeline safety, Asian market potential, the oil sands' compatibility with Canada's climate policies, etc.) and offered top-down explanations to secure public acceptance. The campaign website (https://keepcanadaworking.ca) and video series were built around short, FAQ-style prompts (e.g., "How will expanding the Trans Mountain Pipeline benefit Canadians?", "How will the Trans Mountain Pipeline contribute to Canada's greener energy future?", "How is Trans Mountain part of the national climate plan?"), each addressed by concise talking points in short explainer videos. The campaign website bundled these videos with shareable "facts," carousels showing stakeholder testimonies, and a widget inviting citizens to join "the majority of Canadians who support the Trans Mountain Pipeline expansion" (Fig. 3.1).

The KCW website purposefully copied the structure of public awareness campaigns initiated by citizen groups and non-governmental organizations. KCW's central storyline, namely "TMX is a national project that benefits all Canadians," was narrated as follows: (a) the federal acquisition made all Canadians stakeholders in TMX, resolving disputes between public and private interests; (b) a curated group of TMX supporters was positioned as the majority representing the nation, while dissenting stakeholders (particularly in British Columbia) were portrayed as localized obstacles; (c) considerations of climate change, safety, and public consultation were framed not as constraints but as project merits, portraying

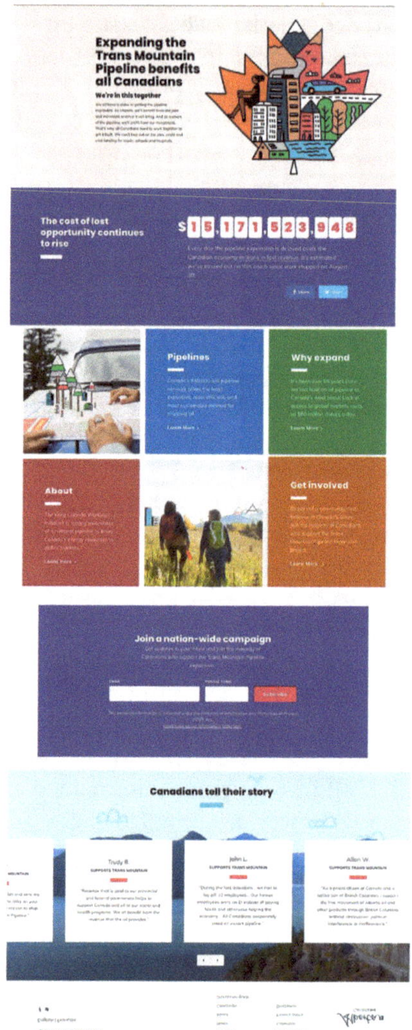

Fig. 3.1 Screenshot of KCW homepage (*Source* Internet Archive, capture of website version 2019 March 08)

oil exports as essential to the energy transition. Underlying the storyline was an ideological move that constructed extractivism as an index of citizenship and solidarity while depicting dissent as anti-Canadian (Gunster et al., 2021; Koch & Perreault, 2019).

What Is "Oil Sands"?

The KCW website largely avoided the term "oil sands." When addressing "what is a pipeline" and "why expand," it spoke instead of "hydrocarbons such as bitumen, refined fuels, and natural gas" being transported by pipelines from one point to another and the importance of TMX for getting higher prices for "our oil" by reaching the global market. Thus, the website treated oil sands, an unconventional fossil fuel, as equivalent to conventional hydrocarbon deposits, and visitors' attention was directed toward infrastructure and transportation logistics.

The information on infrastructure focused on downplaying TMX's scale, presenting it as a prudent upgrade of a legacy system rather than a forthcoming driver of upstream oil sands production. The website's "Route" page emphasized that TMX was "designed to minimize its impact on the Canadian landscape by adding to existing infrastructure that's been in place for 65 years." It supported this claim by noting that TMX would follow 73% of the current pipeline route and 16% along existing utility corridors, such as power lines and highways; only 11% would take new routes. The page's embedded video reinforced these figures using infographics (Fig. 3.2). The project was illustrated as a teal line on a white map of Canada, minimizing the intricate geographic and ecological conditions along the pipeline route. Meanwhile, the video showed no rigs, mines, or tailings, keeping the project's role in incentivizing oil sands production off-screen.

Similarly, on the website's "Marine Safety" page, tanker traffic associated with TMX was depicted as a marginal increment of an already busy, well-regulated port:

> Vancouver is home to Canada's busiest port [...] and about 250 commercial shipping vessels moving into and out of the Port of Vancouver every month. When the Trans Mountain expansion is complete and operational, 34 of those commercial vessels will be oil tankers. That's roughly 1 oil tanker a day.

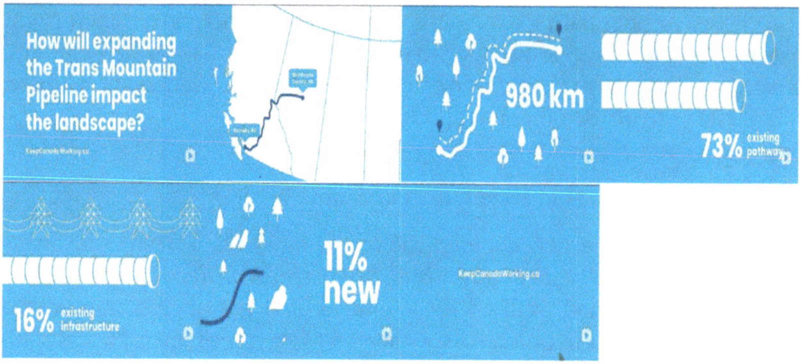

Fig. 3.2 "How will expanding the Trans Mountain pipeline impact the landscape?"

The page also emphasized that "in over 60 years of loading tankers with oil, Trans Mountain has had zero spills from marine operations." The term "marine operations" was employed to set apart tanker transportation from the previously mentioned Westridge 2007 spill. Similar to the visual explanations provided for TMX's route, the page's embedded video (Fig. 3.3) employed pictograms and infographics to mitigate the perception of challenges related to tanker traffic from the Burrard Inlet.

The KCW website presented additional talking points—notably national prosperity and decarbonization—to distract visitors from the oil sands industry. These talking points repositioned the oil sands industry as a revenue generator for decarbonization, rather than a major contributor to greenhouse gas emissions. For example, the "Why Expand" page highlighted that, according to the Alberta government's estimation, the delay of TMX had resulted in a daily loss of approximately $80 million to the Canadian economy, a loss that could have funded essential public services such as health care, schools, roads, public transit, and hospitals. Meanwhile, the "Climate Action" page asserted that TMX would also provide funding for green investments, shift the transportation of oil away from more carbon-intensive methods such as rail or truck, and establish a bridge to a greener future.

Fig. 3.3 What does the Trans Mountain pipeline expansion mean for marine traffic?

Who Are "We" in Relation to Oil Sands?

The KCW website constructed "we" as a collective stakeholder of TMX and invited visitors to join it. Its header image (Fig. 3.4) declared, "Expanding the Trans Mountain Pipeline benefits all Canadians," followed by "We're in this together" and "As owners of the pipeline, we'll profit from our investment." Then, to urge "all Canadians" to join a "nationwide campaign" to get the project built, the homepage featured a live lost-revenue counter, emphasizing that each day of delay in the pipeline expansion costs the Canadian economy millions in lost revenue.

Building on Ottawa's announcement to acquire the Trans Mountain pipeline and its expansion project, the website's "Ownership" page explained why TMX merits the support of the majority of Canadians. The page argued that federal ownership would not only help remove regulatory obstacles in British Columbia but also ensure a financial return to Canada in four ways: (1) profits from both new and existing pipelines, (2) government revenue from the expansion (projected at $46.7 billion over

Fig. 3.4 KCW header image and lost-revenue counter

the next two decades), (3) economic activity stimulated by the project, and (4) increased investor confidence in Canada's business environment. These legitimized TMX as a project in Canada's national interest.

To prevent "we" from being read as an abstract collective, KCW anchored TMX in everyday life by promising "more jobs and a stronger

Fig. 3.5 Expression of nationalism in "How will expanding the Trans Mountain Pipeline make us less reliant on the United States"

economy," funding for "roads, schools, and hospitals," and green investments to assist Canada in transitioning to a clean-energy future. Images of families, nurses, students, and maple-leaf icons in KCW's videos (e.g., Fig. 3.5) symbolized this everyday nationalism, linking ordinary life to energy infrastructure. The website also featured a curated carousel of testimonies from premiers, mayors, business associations, unions, and select Indigenous leaders to demonstrate widespread support for TMX (Fig. 3.6). In short, the "we" KCW constructed was a public that (i) owns TMX, (ii) does the right thing by supporting the project, and (iii) benefits through profits and taxes.

Who Are "They"?

The KCW website identified two composite antagonists. On the "Ownership" page, the principal obstacle was the BC government, which "took on different tactics to delay the project." The federal purchase was presented as the remedy, since "tactics (used by the BC government) to discourage a private company would not work on the federal government, which has special immunities." The United States, a monopolistic

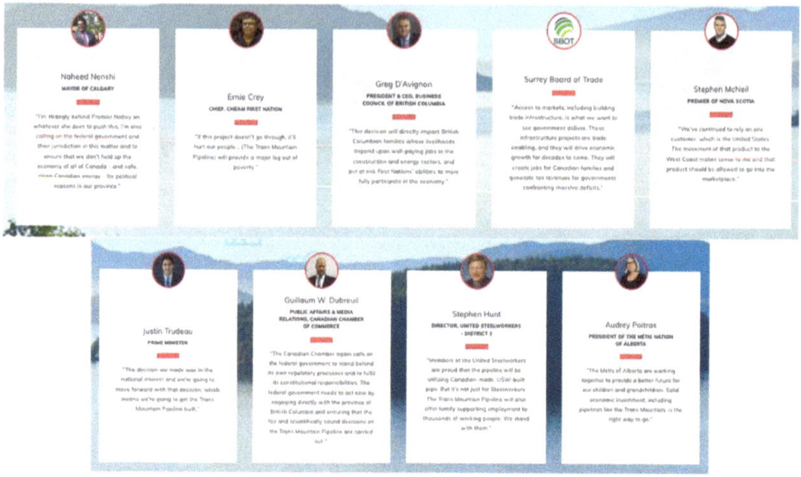

Fig. 3.6 Carousel of testimonials

buyer that drove down the price of oil sands, was the second antagonist. The "Why Expand" page stated that "relying on a single customer (i.e., the United States) for our energy products is problematic for a number of reasons," framing "market access" as an issue of sovereignty and "diversifying export capacity" as a matter of national interest.

Claims of due diligence and delay costs were employed to mitigate procedural opposition. The KCW website stressed that TMX was approved after extensive regulatory and environmental reviews, and that the 157 conditions and recommendations from the National Energy Board (Canada Energy Regulator, 2024; Canadian Press, 2024) confirmed that the project had been duly vetted. Meanwhile, the live "lost-revenue" counter on the home page (Fig. 3.4) equated opposition to foregone public revenue, thereby portraying "their opposition" as a drain on public funds that could have been allocated to hospitals, schools, and roads.

Public concerns regarding TMX's environmental impacts were indirectly addressed through technocratic reassurance rather than public engagement. For example, the "Marine Safety" page underscored a minor increase in marine traffic (one oil tanker per day) and zero spills from marine operations in over 60 years. These data, in conjunction with measures like the Oceans Protection Plan, suggested that risks would be mitigated through expert regulation. In short, the "they" KCW constructed were regional opponents who were overly concerned with risks and abused the regulatory review procedure, as well as a monopoly buyer. The "we," by contrast, were responsible federal owners and expert regulators who would serve the national interest by allowing an already-vetted project to be built.

What Are "We" Going to Do with Oil Sands?

According to KCW, the future of the oil sands sector is based on three interrelated pillars. First and foremost, TMX must be constructed since it will enable increased revenue generation and a reduction in reliance on the United States. Second, the higher revenue, besides funding public programmes and services, serves as a fiscal catalyst for decarbonization. The claim that "oil sands can act as a bridge to a more sustainable future" was reiterated throughout KCW videos and webpages (e.g., Fig. 3.7). Third, the carbon footprint of the oil sands industry will be decreased and made more sustainable through enhanced marine traffic monitoring technologies and procedural due diligence. In sum, these pillars are united by a mobilizing script: Canadians should entrust expert regulation to environmental risks and assist in the construction of federally approved projects as owners. By doing so, they are entitled to improved public services and climate policy that are financed by fossil fuel revenue.

3.4 Assessment: Envisioning Canada as a "Pipeline Nation"

All things considered, KCW made a distinctive rhetorical move by fusing TMX with national belonging. The campaign website's framing of the federal purchase as state ownership and its invitation for all Canadians to join a nationwide campaign to expand the oil sands industry's export capacity transformed a rent-seeking industry into a symbol of solidarity and nation-building. This echoed Barney's (2017) discussion of "pipeline nationalism": Mega energy projects do not merely serve the economy; they materialize the nation, binding territory while providing a medium for rhetoric that invites the public to join the technocratic future set by the fossil fuel sector. Pipeline projects such as TMX have transformed into sites where private accumulation is reinterpreted as serving the public good, while dissent is marginalized as unpatriotic.

Fig. 3.7 Screenshots of "How is Trans Mountain Part of Canada's National Climate Plan"

However, KCW's celebration of state control over the oil sands industry was largely symbolic. Although Ottawa acquired TMX, the upstream complex of oil sands production remained in private, often foreign, hands (Carroll, 2021; Nikiforuk, 2010). KCW discussed the public benefits of TMX at length while neglecting questions about the industry's overall ownership and capital accumulation. Thus, KCW's rhetoric of "we're in this together" and state ownership used nation-building appeals to celebrate a vast collective resource that was available for private appropriation (Barney, 2017). These appeals abstracted corporate interests into "the economy" and bolstered them with tropes of sovereignty and independence.

The utilization of resource nationalism as a campaign strategy resulted in only limited success. The KCW campaign turned out to be short-lived—it was abandoned after the Alberta NDP's defeat in the 2019 provincial election. As of 2025, most campaign materials have been taken down, so I relied on the Wayback Machine (https://web.archive.org/) to retrieve the campaign's webpages. The campaign videos were collected with the help of my colleague Dr. Patrick McCurdy at the University of Ottawa; his "Mediatoil" project traced how the costs and benefits of oil sands development were portrayed in advertisements (see McCurdy, 2018).

During 2018–2019, the legitimacy of TMX was not stabilized by KCW. On August 30, 2018, the Federal Court of Appeal revoked Ottawa's approval, noting that the NEB's review was too flawed to back up the decision (Gibson & Reiger, 2018). This forced the federal government to undertake a new review and consultation process for TMX, ultimately re-approving it on June 18, 2019, with a commitment to reinvest federal revenues from the project into clean energy and to revert the expanded asset to private ownership (Connolly & Ferreras, 2019). Meanwhile, public opinion never coalesced into the consensual "we" that KCW constructed: polls conducted in mid-2018 found Canadians divided on the federal purchase and on TMX more broadly (Angus Reid Institute, 2018; Ipsos, 2018).

Therefore, KCW's information-deficit-based messaging and technocratic assurances did not settle the fundamental disputes over jurisdiction, cumulative climate impacts, tanker risk, and Indigenous sovereignty. This demonstrates the limits of promoting oil sands as a national industry: the campaign's nationalist claims obscured persistent regional cleavages—especially in coastal B.C.—and the fact that those most exposed to risk (Indigenous nations, marine ecosystems, urban port communities) are least empowered in revenue allocation. One element, however, proves durable. KCW helped cement the coupling of "energy" and citizenship—ownership rhetoric, welfare promises, and climate-funding claims—that subsequent oil-sands advocates would build upon. In the years that followed, the oil sands industry and its supporters increasingly cast their critics as anti-Canadian, extending KCW's collective-shareholder imaginary into a more punitive politics of belonging. I will detail this trend in the next chapter.

References

Anderson, D. (2018, April 8). Arrests to continue as Kinder Morgan protests heat up. *The Tyee*. https://thetyee.ca/News/2018/04/04/Kinder-Morgan-Protest-Arrests

Angus Reid Institute. (2018, June 18). *Good Investment or Bad Precedent? Canadians pick sides on TransMountain pipeline purchase*. https://angusreid.org/federal-transmountain-purchase/

Arbatli, E. (2018). Resource nationalism revisited: A new conceptualization in light of changing actors and strategies in the oil industry. *Energy Research & Social Science, 40*, 101–108. https://doi.org/10.1016/j.erss.2017.11.030

Barney, D. (2017). Who we are and what we do: Canada as a pipeline nation. In S. Wilson, A. Carlson, & I. Szeman (Eds.), *Petrocultures: Oil, politics, culture* (pp. 86–126). McGill-Queen's University Press.

Canada Energy Regulator. (2024). Trans Mountain Expansion: Project Background. https://www.cer-rec.gc.ca/en/applications-hearings/view-applications-projects/trans-mountain-expansion/project-background.html

Canadian Press. (2024, January 30). A timeline of the Trans Mountain pipeline expansion's major milestones and setbacks. *Financial Post*. https://financialpost.com/pmn/business-pmn/a-timeline-of-the-trans-mountain-pipeline-expansions-major-milestones-and-setbacks

Carroll, W. K. (ed.) (2021). *Regime of obstruction: how corporate power blocks energy democracy*. Athabasca University Press.

Chagnon, C. W., Durante, F., Gills, B. K., Hagolani-Albov, S. E., Hokkanen, S., Kangasluoma, S. M. J., Konttinen, H., Kröger, M., LaFleur, W., Ollinaho, O., & Vuola, M. P. S. (2022). From extractivism to global extractivism: The evolution of an organizing concept. *The Journal of Peasant Studies, 49*(4), 760–792. https://doi.org/10.1080/03066150.2022.2069015

Childs, J. (2016). Geography and resource nationalism: A critical review and reframing. *The Extractive Industries and Society, 3*(2), 539–546. https://doi.org/10.1016/j.exis.2016.02.006

Coast Protectors. (n.d.). Trans Mountain is a major climate problem. https://www.coastprotectors.ca/trans_mountain_climate_cost. Accessed August 28, 2025.

Connolly, A., & Ferreras, J. (2019, June 19). Trans Mountain pipeline expansion gets green light to proceed. *Global News*. https://globalnews.ca/news/5403711/trans-mountain-pipeline-approved/

Conversi, D. (2020). The ultimate challenge: Nationalism and climate change. *Nationalities Papers, 48*(4), 625–636. https://doi.org/10.1017/nps.2020.18

Cypher, J. M. (2010). South America's commodities boom: Developmental opportunity or path dependent reversion? *Canadian Journal of Development*

Studies / Revue Canadienne D'études Du Développement, 30(3–4), 635–662. https://doi.org/10.1080/02255189.2010.9669319

Gibson, J., & Reiger, S. (2018, August 30). Trans Mountain pipeline ruling strikes 'devastating' blow to Canada's global reputation, energy sector warns. *CBC News*. https://www.cbc.ca/news/canada/calgary/kinder-morgan-business-reaction-industry-court-ruling-capp-oil-pipeline-trans-mountain-1.4804934

Gunster, S., Fleet, D., & Neubauer, R. (2021). Challenging petro-nationalism: Another Canada is possible? *Journal of Canadian Studies, 55*(1), 57–87. https://doi.org/10.3138/jcs.2019-0033

Gunster, S., & Saurette, P. (2014). Storylines in the sands: News, narrative, and ideology in the *Calgary Herald*. *Canadian Journal of Communication, 39*(3), 1–27. https://doi.org/10.22230/cjc.2014v39n3a2830

Gunton, T. (2020). Staple theory and the new staple boom. In J. Stanford (Ed.), *The staple theory @ 50: Reflections on the lasting significance of Mel Watkins' "a staple theory of economic growth"* (pp. 43–52). Canadian Centre for Policy Alternatives. https://policyalternatives.ca/publications/reports/staple-theory-50

Haslam, P., & Heidrich, P. (eds.) (2016). *The political economy of natural resources and development: From neoliberalism to resource nationalism*. Routledge.

Henriques, I., & Böhm, S. (2022). The perils of ecologically unequal exchange: Contesting rare-earth mining in Greenland. *Journal of Cleaner Production, 349*, Article 131378. https://doi.org/10.1016/j.jclepro.2022.131378

Hoberg, G. (2016). Pipelines and the politics of structure: A case study of the Trans Mountain Pipeline. Paper presented at the 2016 Annual Meeting of the Canadian Political Science Association. https://cpsa-acsp.ca/documents/conference/2016/Hoberg.pdf

Home, A. (2024, March 12). Tin supply trapped in resource nationalism squeeze. *Reuters*. https://www.reuters.com/markets/commodities/tin-supply-trapped-resource-nationalism-squeeze-2024-03-12/

Huber, M. T. (2013). *Lifeblood: Oil, freedom, and the forces of capital*. University of Minnesota Press.

Innis, H. A. (1970). *The fur trade in Canada: An introduction to Canadian economic history*. (Rev. ed.). University of Toronto Press.

International Energy Agency. (2024, March 8). *(Alberta) Climate Leadership Plan 2015*. https://www.iea.org/policies/8583-alberta-climate-leadership-plan

Ipsos. (2018, July 19). *Canadians evenly divided on fed's purchase of Trans Mountain Pipeline: 42% support, 42% oppose, 16% don't know*. https://www.ipsos.com/en-ca/news-polls/Global-News-Pipeline-Poll-July-18-2018

Jaspal, R., Nerlich, B., & Lemańcyzk, S. (2014). Fracking in the Polish press: Geopolitics and national identity. *Energy Policy, 74*, 253–261. https://doi.org/10.1016/j.enpol.2014.09.007

Karl, T. L. (1997). *The paradox of plenty: Oil booms and petrostates*. University of California Press.

Kennedy, E. (2014). From petro-states to 'new realities': Perspectives on the geographies of oil. *Geography Compass, 8*(4), 262–276. https://doi.org/10.1111/gec3.12127

Koch, N., & Perreault, T. (2019). Resource nationalism. *Progress in Human Geography, 43*(4), 611–631. https://doi.org/10.1177/0309132518781497

Kraushaar-Friesen, N., & Busch, H. (2020). Of pipe dreams and fossil fools: Advancing Canadian fossil fuel hegemony through the Trans Mountain pipeline. *Energy Research & Social Science, 69*, Article 101695. https://doi.org/10.1016/j.erss.2020.101695

Laastad, S. G. (2024). Leaving oil in the ground: Ecuador's Yasuní-ITT initiative and spatial strategies for supply-side climate solutions. *Environment and Planning. A, 56*(1), 172–189. https://doi.org/10.1177/0308518X231184876

LeQuesne, T. (2019). Petro-hegemony and the matrix of resistance: What can Standing Rock's water protectors teach us about organizing for climate justice in the United States? *Environmental Sociology, 5*(2), 188–206. https://doi.org/10.1080/23251042.2018.1541953

Lis, A. (2018). Co-production of the shale gas publics in Poland and the negotiation of the state citizens relations. *The Extractive Industries and Society, 5*(4), 673–681. https://doi.org/10.1016/j.exis.2018.07.005

Materka, E. (2011). End of transition? Expropriation, resource nationalism, fuzzy research, and corruption of environmental institutions in the making of the shale gas revolution in northern Poland. *Debatte, 19*(3), 599–631. https://doi.org/10.1080/0965156X.2012.681919

McCurdy, P. (2018). From the natural to the manmade environment: The shifting advertising practices of Canada's oil sands industry. *Canadian Journal of Communication, 43*(1), 33–52. https://doi.org/10.22230/cjc.2018v43n1a3315

Mitchell, T. (2011). *Carbon democracy: Political power in the age of oil* (1st ed.). Verso.

Moreau, J. (2015, May 20). Burnaby Mayor Derek Corrigan ready to end career with pipeline arrest. *Vancouver Sun*. https://vancouversun.com/news/metro/burnaby-mayor-derek-corrigan-ready-to-end-career-with-pipeline-arrest

Nikiforuk, A. (2010). *Tar sands: Dirty oil and the future of a continent* (Revised ed.). Greystone Books.

Pryke, S. (2017). Explaining resource nationalism. *Global Policy, 8*(4), 474–482. https://doi.org/10.1111/1758-5899.12503

Quiliconi, Q., & Vasco, P. R. (2021, September 20). Chinese mining and Indigenous resistance in Ecuador. Carnegie Endowment for International Peace. https://carnegieendowment.org/research/2021/09/chinese-mining-and-indigenous-resistance-in-ecuador

Radetzki, M. (2006). The anatomy of three commodity booms. *Resources Policy, 31*(1), 56–64. https://doi.org/10.1016/j.resourpol.2006.06.003

Raincoast. (2019, November 5). *Trans Mountain Expansion & Southern Resident killer whales: Project background*. https://www.raincoast.org/2019/11/trans-mountain-expansion-project-background

Riofrancos, T. (2020). Resource radicals: From petro-nationalism to post-extractivism in Ecuador. *Duke University Press*. https://doi.org/10.1515/9781478012122

Short, J. W. (2015, May 11). *Fate and effect of oil spills from the Trans Mountain Expansion project in Burrard Inlet and the Fraser River Estuary*. JWS Consulting LLC. https://twnsacredtrust.ca/wp-content/uploads/2015/05/TWN-Assessment-Appendix-3.pdf

Stanford, J. (ed.) (2020). *The staple theory @ 50: Reflections on the lasting significance of Mel Watkins' "a staple theory of economic growth"*. Canadian Centre for Policy Alternatives. https://policyalternatives.ca/publications/reports/staple-theory-50

Stewart, I. G., & Harding, M. E. (2023). One pipeline and two impact assessments: Coproduction, legal pluralism, and the Trans Mountain Expansion project. *Science, Technology, & Human Values, 48*(3), 525–551. https://doi.org/10.1177/01622439211057309

Sturgis, P., & Allum, N. (2004). Science in society: Re-evaluating the deficit model of public attitudes. *Public Understanding of Science, 13*(1), 55–74. https://doi.org/10.1177/0963662504042690

Szeman, I. (2013). How to know about oil: Energy epistemologies and political futures. *Journal of Canadian Studies, 47*(3), 145–168. https://doi.org/10.1353/jcs.2013.0020

Taft, K. (2017). *Oil's deep state: how the petroleum industry undermines democracy and stops action on global warming in Alberta, and in Ottawa*. James Lorimer & Company Ltd.

Trans Mountain Corporation. (n.d.-a). Past Project: TMEP. https://www.transmountain.com/past-project-tmep (Accessed: August 28, 2025)

Trans Mountain Corporation. (n.d.-b). Westridge 2007 Spill. https://www.transmountain.com/westridge-2007-spill (Accessed: August 28, 2025)

Trans Mountain pipeline expansion not supported by B.C. Government. (2016, January 11). *CBC News*. https://www.cbc.ca/news/canada/british-columbia/transmountain-b-c-government-kindermorgan-1.3398689

Veltmeyer, H., & Bowles, P. (2014). Extractivist resistance: The case of the Enbridge oil pipeline project in Northern British Columbia. *The Extractive Industries and Society, 1*(1), 59–68.

Watkins, M. H. (1963). A staple theory of economic growth. *The Canadian Journal of Economics and Political Science, 29*(2), 141–158. https://doi.org/10.2307/139461

White, R., & Kramer, R. C. (2015). Critical criminology and the struggle against climate change ecocide. *Critical Criminology, 23*(4), 383–399. https://doi.org/10.1007/s10612-015-9292-5

Wilson, J. D. (2015). Understanding resource nationalism: Economic dynamics and political institutions. *Contemporary Politics, 21*(4), 1–18. https://doi.org/10.1080/13569775.2015.1013293

Wilson, S., Carlson, A., & Szeman, I. (Eds.). (2017). *Petrocultures: Oil, politics, culture* (1st ed.). McGill-Queen's University Press. https://doi.org/10.1515/9780773550391

Yusufali, S., & Pratt, L. R. (2009, November 16). Petro-Canada. *The Canadian Encyclopedia*. https://thecanadianencyclopedia.ca/en/article/petro-canada

Żuk, P., Conversi, D., & Żuk, P. (2024). Conceptualising energy nationalism in the context of climate change: framework and review. *Frontiers in Energy Research, 12*, Article 1349581. https://doi.org/10.3389/fenrg.2024.1349581

CHAPTER 4

"Public Inquiry into Anti-Alberta Energy Campaigns": The Rise of Extractive Populism

Abstract Focusing on the 2019–2021 "Public Inquiry into Anti-Alberta Energy Campaigns," the chapter traces a shift from "national interest" messaging to enemy construction and smear tactics that depict opponents as foreign-funded radicals. It conceptualizes this shift as extractive populism: a petro-bloc politics that moralizes markets, securitizes dissent, and operationalizes "the people" through platform-mediated mobilization and state-adjacent branding, while channelling solutions toward communication strategy rather than regulatory change. The chapter demonstrates how this repertoire retools Polanyi's double movement to defend markets, infrastructures, and export corridors as objects of protection.

Keywords Double movement · Extractive populism · Neoliberal populism · Petro-bloc · Platform mobilization · Radicalization

This chapter focuses on the developments of the Trans Mountain Pipeline controversy after the 2019 Alberta general election, in which the Conservatives regained the majority. Compared with the "Keep Canada Working" campaign's emphasis on national interest, the current Conservative government of Alberta has shifted the focus of its pro-oil sands

S. Chen, *Petro-Rhetoric in Canada*, Palgrave Studies in Media and Environmental Communication,
https://doi.org/10.1007/978-3-032-12964-2_4

narratives to portray opponents of the oil sands industry as environmental radicals supported by foreign interests. This shift toward smear tactics reflects the increasing radicalization of Canada's "petro-bloc," whose members embrace alternative facts in order to refute the broad public consensus regarding the phase-out of fossil fuels. This chapter explains, via a detailed critical analysis of the "Public Inquiry into Anti-Alberta Energy Campaigns" (2019–2021), the factors that contribute to the emergence of "extractive populism" in Canada and the discursive features of this trend.

4.1 The Rise of Extractive Populism in Canada

Karl Polanyi's (2001 [1957]) concept of "double movement" posits that capitalist markets drive the disembedding of the economy from social constraints, permeating all aspects of life with market logic; concurrently, this triggers a counter-movement within society aimed at protecting against the destabilizing effects of the market. This push-and-pull has been a fundamental factor in the rise of traditional populist movements, which function as a reactionary force against neoliberalism (Lim, 2023). Key characteristics of traditional populism comprise: (1) a moral division between virtuous "people" and corrupt "elite," (2) pledges to safeguard or reinstate the populace's fair share of national wealth, (3) a collective identity of "the people," and (4) a focus on direct democracy and mass mobilization against the elite (Bekmen & Alp Özden, 2022; Rossi, 2025).

Yet, as demonstrated by the rise of right-wing populist parties and movements in recent years (e.g., Trumpism in the United States, Bolsonarismo in Brazil, the Five Star Movement in Italy, etc.), contemporary populism frequently functions within the framework of neoliberalism, rather than opposing it. This trend signifies the increasing impact of "neoliberal populism," wherein neoliberalism serves as the foundational ideology that provides the norms—market rationality, anti-statism, and competitive individualism—through which appeals to "the people" are articulated (Bekmen & Alp Özden, 2022; Konings, 2012; Queiroz, 2021; Rossi, 2025; Scheiring, 2022). Neoliberal populism preserves core neoliberal commitments (e.g., market freedom, fiscal discipline, competitiveness, etc.) while using nationalist and anti-elitist rhetoric to claim these are the authentic will of "ordinary people."

One classic example of this synthesis in the United States was Reaganite discourse, which claimed to defend the interests of hard-working "Americans" (craftsmen, farmers, shopkeepers, etc.) against a parasitic bureaucracy, thereby presenting tax cuts and deregulation as popular emancipation rather than upward redistribution (Guardino, 2018). In this case, neoliberalism's political project was legitimized by affective and ethical frames—discipline, self-help, tough love, thrift, and patriotism—that converted public anxiety about the economy into demands for austerity imposed on others. Similarly, American right-wing populist movements have demonstrated remarkable endurance in the wake of the 2008 financial crisis. Konings (2012) argues that this endurance results from the affective and moral appeals that are embedded in their discourses. By evoking the ethos celebrated by Reaganite discourse, these movements resonate emotionally with the public, motivating individuals to not only endure austerity but also to demand it from others.

Beyond the United States, neoliberal populism also gains traction in countries such as Brazil, Italy, Hungary, Thailand, and Turkey, where market-based inclusion and social assistance are offered amid socio-economic crises (Bekmen & Alp Özden, 2022; Rossi, 2025; Scheiring, 2022). In these contexts, citizenship becomes increasingly linked to consumption and indebtedness rather than to political and social rights. This shift weakens collectivity, fosters the rise of charismatic leaders, and increases the risk of a slide toward authoritarianism.

Therefore, neoliberal populism rewrites the dynamics of "double movements" by promising moral purification through marketized autonomy, fusing "resentment" with redemption, and recoding protection as disciplined, self-reliant freedom. It gains momentum by channelling public anxiety into a moral framework that frames austerity as a virtue and portrays progressive policies as hypocritical and out of touch.

Neoliberal populism's most critical discursive feature is its use of "the people" solely as a rhetorical container for entrepreneurial individuals. It prioritizes the unbridled pursuit of private gain and casts welfare recipients, migrants, and bureaucrats as unproductive threats to the nation's moral economy (Bekmen & Alp Özden, 2022; Rossi, 2025). This yields a paradox: the collective is symbolically exalted, while policy solutions intensify competitive individualism and extend market logics into citizenship itself. By turning citizens into a de-collectivized mass, neoliberal populism pursues a class-hegemony project that consolidates power through selective inclusion: expanding market-mediated assistance and consumer citizenship while weakening collective rights and paving paths to authoritarian rule when crises unfold.

Another major feature of neoliberal populism is its reliance on social media platforms, which mediate the voices of an informal, often uninformed mass (Gerbaudo, 2018; Rossi, 2025). Neoliberal populism rejects traditional political elites but replaces them with technocratic or platform-based leadership, with influencers promoting market-friendly, efficiency-based rhetoric and diverting public attention from substantive critiques of the neoliberal order. Under neoliberal populism, passive "users" engaged primarily through online participation, and the combination of superficial participatory mechanisms and centralized decision-making (by right-wing influencers and politicians) results in an empty concept of "the people" that can be easily redefined by the right's exclusionary politics.

In short, neoliberal populism is characterized by the combination of populist rhetoric and neoliberal governance logic, resulting in an unexpected relationship in which neoliberalism serves as a host ideology for a hollowed-out form of populism. This relationship is also evident in instances of "extractive populism": once anti-elitism and moralized market freedom are in place, resource nationalism can be presented as the populace's defence against cosmopolitan "enemies," while the policy payload remains firmly neoliberal.

Concerning the relationship between populism and environmental issues, previous research has identified four overlapping strands of "environmental populism": (1) right-wing populism and climate denial, which delegitimize climate science and decarbonization; (2) resource or extractive populism, in which leaders mobilize extractive rents to build mass support; (3) eco-populism, which refers to grassroots, place-based struggles that pit communities against exploitative elites; and (4) populist narratives within environmental movements themselves (Bosworth, 2020; Marquardt & Lederer, 2022; Matsen et al., 2016; Ofstehage et al., 2022; Sconfienza, 2022). In this chapter, I focus on extractive populism, which drives environmental conflicts to revolve around a moralized defence of "the people," their land, and national sovereignty against presumed outsiders—scientific and bureaucratic "elites," urban environmentalists, and immigrants—while consolidating executive power to direct resource policy. Additionally, extractive populism, through conservative media and identity politics, fuses anti-elitism and nationalism with climate scepticism.

According to Matsen and his colleagues (2016), resource-dependent economies—from developing countries like Iran and Venezuela to more developed ones like Russia and Norway—tend to reinforce the convergence of extractivism and populism. This is because politicians of all types, motivated to remain in office, lean toward short-term overprovision of goods and services to secure political support; in turn, voters reward this unsustainable spending even when they realize extraction-based goods and services cannot be sustained over time.

The convergence of extractivism and populism, however, could encounter resistance from eco-populism. Protests against pipeline projects across North America, for example, have formed an "unlikely alliance" of Indigenous nations, ranchers, farmers, and climate groups whose populist counter-narratives opposed fossil capital and distant regulators (Hess, 2023; Hoberg, 2021; Neubauer & Gunster, 2019). Campaigns by this alliance helped catalyse a broader anti-pipeline movement that contributed to delays and even cancellations of major projects such as Keystone XL and Northern Gateway. Parallel transnational mobilizations, such as peasant and civil-society coalitions against extractivism in Latin America (as illustrated in the case of Ecuador discussed in Chapter 3), demonstrate how rural constituencies challenge authoritarian, market-oriented land and resource schemes (Riofrancos, 2020).

It is important to note that although eco-populism can foster public resistance through narratives of place and identity, extractive populism can achieve similar public mobilization through narratives that combine resource nationalism with classed and gendered imaginaries about regional identity. For example, in Minnesota's Iron Range, a historical mining district in the state's northeastern corner, pro-mining narratives invoked the "insider versus outsider" dichotomy, anger at environmental regulation, nostalgia for mining as a "way of life," and the claim that domestic extraction secures the national economy (Kojola, 2019). Such narratives made extractive expansion feel like communal defence rather than merely economic policy and helped shift a historically Democratic region toward Trump.

Taken together, extractive populism moralizes resource extraction as a means of safeguarding public interests while simultaneously transferring power over land, labour, and environmental policy to governments, corporations, and markets. In Canada, this ideology has facilitated the development of narratives that portray oil sands expansion as popular sovereignty in opposition to "environmental radicals" and "foreign adversaries."

Specifically, the convergence of resource extraction and regional identity in Canada predates social media. In the late 1960s, the development of large-scale oil sands projects coincided with a shift to mediatized, image-driven politics, which McCurdy (2023) calls "synthetic politics." In this context, rhetoric, visuals, and policies—produced with media logics in mind—laid the ideological foundation for later forms of petro-hegemony.

During the 2000s–2010s, a discursive triad consolidated: celebration of oil sands as making Canada an "energy superpower" and enhancing its energy security, technocratic appeals to scientific management of oil sands' environmental harm, and the moral framing of oil sands as "ethical oil" (Kinder, 2024; Kuteleva & Leifso, 2020; Laurie, 2019). This triad linked fossil fuel development to the common good while discrediting environmental criticisms that exceeded regulatory thresholds.

Meanwhile, as oil sands pipelines encountered regulatory delays and public resistance, conservative media outlets—exemplified by the *National Post* and the *Calgary Herald*—amplified a "besieged industry" storyline that portrayed Alberta as under attack by powerful outsiders, framed oil sands as a public good, and downplayed unequal distributions of risk and benefits (Chen, 2019; Gunster & Saurette, 2014). Joe Oliver, the federal Natural Resources Minister of the Stephen Harper government (2006–2015), further bolstered this storyline by cautioning against "foreign-funded radicals" and "jet-setting celebrities" who posed a threat to Canada's energy future. This represented an archetypal expression of extractive populism.

The "besieged industry" storyline is neither spontaneous nor purely state-led. Its prevalence has been driven by the consolidation of the Canadian "petrobloc": "a decentralized yet interlocked constellation of state, civil society, and corporate actors jointly dedicated to tar sands expansion" (Neubauer, 2018, p. 249). The Canadian petro-bloc amplifies pro-oil sands arguments across media and public debate by supplying cultural capital and "interfield" legitimacy—technocratic authority alongside apparent grassroots voices. The block model explains why pipeline projects in Canada maintain high public profiles, despite the fact that the oil and gas industry accounts for a modest proportion of national employment and exports in comparison with classic petro-states.

In response to social media's transformation of publicity, industry advocacy has adopted "subsidized publics" strategies (Gunster et al., 2021). Elites provide prefabricated content, training, and mobilization opportunities to supporters to lower participation costs. The "Canada's

Energy Citizens" campaign exemplifies this shift, supplying prepackaged media for circulation through personal networks—an approach that relies on trust among friends, family, and neighbours. According to Wood (2018), the campaign performs ideological work by normalizing pro-industry voices and reframing public relations labour as civic action. In doing so, it makes supportive citizens in Canada's petro-bloc more visible and vocal.

On social media, coordinated industry efforts simulate, amplify, and organize grassroots sentiment at scale, creating a false sense of popular support for the oil sands (Gunster et al., 2021; Neubauer et al., 2023). Kinder (2024) calls this phenomenon "petroturfing," tracing its origins to debates about "ethical oil" in the early 2010s and showing how pro-oil actors portray themselves as marginalized while promoting extraction as national destiny. Among popular platforms, Facebook pages—because they allow key accounts (for example, official campaign pages) to curate, reframe, and bridge frames—play a central role in converting diffuse online voices into a coherent extractive-populist storyline.

Neubauer and his colleagues (2023) name the coordinating role performed by pro-oil sands Facebook pages "connective leadership," which links "Canadian Energy" to "the people" and galvanizes opposition against perceived enemies—foreign funders, cosmopolitan elites, and overreaching regulators—often via direct calls to vote, sign petitions, or pressure officials. Their analysis identified common frames (attack opponents; public/national interest; "ethical oil"; Indigenous benefit) and frequent mobilization appeals, with posts targeting perceived enemies eliciting emotional responses from supporters. One illustrative case of the current platform-native form of extractive populism is the "Oil Sands Strong" campaign. Tian and Ge (2025) conducted a multimodal discourse analysis of its Facebook posts and found a stylized self–other binary (patriotic "Canadian" oil and workers vs. foreign producers, renewables, and urban environmentalists), fused with xenophobic cues and argumentative topoi of reality, threat, advantage, and history to legitimize extraction and portray dissent as un-Canadian.

In sum, the progression from ethical oil to petroturfing signifies the evolution of extractive populism in Canada: a cross-sector bloc symbolically nationalizes the oil sands industry; pro-industry media voices and connective leadership transform the symbolism into everyday narratives; and subsidized publics and influencer pages convert identity work into political mobilization. The outcome is not simply pro-industry rhetoric but a resilient moral economy wherein supporting extraction is cast as defending “the people,” the nation, and common sense. Below I will further elucidate how the Alberta government supplies ideological backing to extractive populism by analyzing the “Public Inquiry into Anti-Alberta Energy Campaigns.“

4.2 The Trans Mountain Pipeline Controversy, 2019–2021

The Trans Mountain Pipeline Expansion (TMX) project transitioned from legal uncertainty to active construction between 2019 and 2021. However, the oil sands industry faced substantial obstacles as a result of the COVID-19 pandemic. As mentioned in Chapter 3, the Federal Court of Appeal revoked the federal government’s approval of TMX on August 30, 2018, necessitating a new review and consultation process (Canadian Press, 2024). The federal government reapproved the project on June 18, 2019. The Federal Court of Appeal and the Supreme Court of Canada dismissed subsequent litigation by several First Nations, providing the project with legal certainty moving forward (Ecojustice, n.d.).

Nevertheless, as the Canadian economy confronted unparalleled difficulties during the COVID-19 pandemic, both Alberta’s oil sands industry and TMX experienced escalating costs and a record decline in oil demand. One iconic event illustrating the pandemic’s devastating impact on the oil and gas sector occurred on 20 April 2020, when US crude fell to a negative value for the first time in history as stockpiles overwhelmed storage facilities (Ambrose, 2020). In the context of widespread layoffs, business closures, and travel restrictions due to the pandemic, Alberta experienced an economic stagnation and diminished oil demand, resulting in a $6.8 billion deficit in 2020 (Riley, 2020). Meanwhile, the construction of TMX, deemed essential work by the Canadian government, continued amid lockdown measures, but this caused a slowdown in progress and additional expenses for pandemic safety protocols. In 2020, TMX disclosed $12.5 million in expenses associated with COVID-19 (Thurton, 2020). This, coupled with numerous safety incidents (e.g., Canadian Press, 2021), fuelled enduring public scepticism regarding TMX’s capacity to adhere to its original budget and to operate as safely as its proponents promised.

The Alberta United Conservative Party (UCP) secured victory in the provincial election on April 16, 2019. It subsequently wrapped up the

"Keep Canada Working" campaign and implemented "fight back" strategies aimed at outsiders who it labelled as conducting an information war against the province's energy interests (Klaszus, 2019). Two pillar initiatives were the result of UCP's aggressive approach. First, on 11 December 2019, it established the Canadian Energy Centre (CEC), the provincial "energy war room" (Anderson, 2019). With an annual budget of $30 million, the CEC's primary objective is to advocate for Alberta oil and gas and to counteract critics. For example, in 2021, CEC garnered international attention by criticizing the Netflix children's film Bigfoot Family as "anti-oil propaganda" (Cecco, 2021).

Second, the UCP government initiated the "Public Inquiry into Anti-Alberta Energy Campaigns" on July 4, 2019, appointing forensic accountant Steve Allan as commissioner to investigate alleged foreign-funded efforts to landlock Alberta's oil and gas (Clancy, 2019). After multiple deadline extensions, the inquiry's 657-page final report was released to the public on October 21, 2021. The report determined that, between 2003 and 2019, foreign funding for "Canadian-based environmental initiatives" amounted to $1.28 billion, with only $54.1 million allocated for what it terms "anti-Alberta resource development activity" (Bellefontaine, 2021). The report also found no evidence of wrongdoing and emphasized that market forces—prices, technology, and geopolitics—primarily explained the oil sands industry's investment decline. Although critics argued the inquiry cost taxpayers $3.5 million and amounted to a witch hunt, the UCP government maintained that it provided evidence of how Albertans were harmed by foreign-funded campaigns. Given the report's pillar role in Alberta's "fight back" strategies, an in-depth analysis will shed light on key ideological underpinnings of the UCP government's embrace of extractive populism.

4.3 Public Inquiry into Anti-Alberta Energy Campaigns

The final report of the Public Inquiry into Anti-Alberta Energy Campaigns (hereafter the "Inquiry Report," Allan, 2021) advanced a narrative that embodied key arguments of extractive populism: (1) resource-driven development is presented as intrinsically beneficial; (2) extractivism is legitimized through discourses of environmental stewardship and technological innovation; (3) opposition is characterized as an externally funded threat to the public good; and (4) counter-mobilization and a rebranding of Canadian energy are proposed as solutions.

The Inquiry Report cited evidence to make the case that anti-oil sands campaigns can influence capital investment decisions and impede extraction. It constructed a virtuous "we" (Albertans/Canadians who value prosperity, ingenuity, and responsible environmental stewardship) and a less virtuous "they" (transnationally networked campaigners) whose actions, while lawful, are normatively misaligned with the national community's economic horizon. This dichotomy associated national virtue with continued oil sands development and identified political impediments as stemming from communication and coordination deficiencies rather than from the sector's environmental contradictions. Rather than advocating regulatory reform, the report proposed a communications-driven initiative to improve public "energy literacy" and to "mitigate polarization," ultimately aiming to make oil sands compatible with national pride and incremental decarbonization and to unite Canadians in support of oil sands development.

The proposed actions reflected Canada's petro-bloc dynamics—elite orchestration of "subsidized publics" and connective leadership that translate corporate agendas into popular will—wherein industry-aligned actors supply frames, content, and calls to action to make support visible and participatory (Gunster et al., 2021; Neubauer et al., 2023). These actions were also consistent with the logic of neoliberal populism: rhetorically exalting the collective while advancing policy instruments—branding, measurement, governance—that stabilize market-centric development and shift structural debate onto communication and compliance.

What Is "Oil Sands"?

The Inquiry Report implicitly defined "oil sands" through a developmentalist imaginary that links hydrocarbons to national purpose and technological modernity. It called for a "long-term strategy" based on a vision of "[Canada] being a global leader in lower-carbon energy," with Alberta "capitalizing" on its advantages rather than "leaving them in the ground." Oil sands was further embedded within an ecosystem of innovation, with pro-oil sands organizations such as Canada's Oil Sands Innovation Alliance, Clean Resource Innovation Network, and government affiliates such as Alberta Innovates and Emissions Reduction Alberta cited as evidence that Alberta's oil sands industry is leading clean-technology investment both nationally and internationally.

"Energy literacy," a pivotal concept emphasized in the report, was framed as a civic obligation: citizens ought to possess comprehensive, trustworthy information and understand the consequences of policy decisions so that development appears both rational and responsible. In short, "oil sands" were discursively constructed as a national industry whose expansion, supported by innovations in clean technology, is compatible with—and even instrumental to—climate stewardship.

Who Are "We" in Relation to Oil Sands?

The Inquiry Report constructed a broad, consensual "we" made up of ordinary citizens, business councils, workers, and Indigenous communities, united by a desire for economic growth and pragmatism. Citing surveys conducted by the University of Ottawa's Positive Energy Centre, it claimed that:

Canadians value the Canadian oil and gas sector and support environmentally responsible development of the sector; they want access to reliable energy; they do not want to pay a lot more, and they do not want to compromise their lifestyle (Allan, 2021, p. 641).

This public support presented an excellent opportunity to begin a public conversation regarding the development and exportation of oil sands. Furthermore, the Business Council of Alberta was portrayed as "well positioned" to provide leadership, alongside industry and environmental organizations, to address public polarization around oil sands. This framing naturalized corporate-civil partnership as democratic common sense, aligning with the connective, elite-brokered forms of participation documented in analyses of Canada's petro-public (Neubauer et al., 2023).

Who Are "They"?

"They" were specified through the Inquiry Report's primary focus: "anti-Alberta energy campaigns," defined as efforts aimed at obstructing or hindering the advancement of Alberta's oil and gas sector, frequently originating from foreign entities. The report quantified at least $54.1 million in "foreign funding" specifically designated for "anti-Alberta resource development activities" and stated that this amount was likely a significant underestimation. Yet, after cataloguing activism—from petitions to protests—the commissioner concluded that:

To be very clear, I have not found any suggestions of wrongdoing on the part of any individual or organization. No individual or organization, in my view, has done anything illegal. Indeed, they have exercised their rights of free speech (Allan, 2021, p. 596).

Thus, the report presented a paradoxical othering: "they" were depicted as foreign-funded, organized, and politically adept, while their methods were characterized as legitimate civil-society advocacy. They were problematized more for their efficacy in influencing public opinion and investment choices regarding oil sands than for any illegality.

What Are "We" Going to Do with Oil Sands?

The Inquiry Report, responding to how to counter anti-oil sands activism, proposed a marketing-governance initiative focused on narrative reconstruction and market repositioning. It proposed the development of a "long-term strategy" centred on "global leader" branding, the improvement of "energy literacy," and the mobilization of a cross-sector coalition to eliminate silos and reduce polarization. The ultimate goal of these measures was clearly supply-side: Canada can and must augment its contribution to climate goals by developing and exporting its energy resources, rather than leaving them unexploited. It should also utilize its resource advantages to achieve greater prosperity. In short, the action plan integrated narratives concerning identity, pedagogy, and competitiveness, redirecting focus from the environmental risks of extraction to communication strategies and coalition management.

4.4 Assessment: The Radicalization of Canada's Petro-Bloc

This chapter has traced the evolution of petro-rhetoric in Canada—from earlier ethical oil to a more combative extractive populism that fuses nation, industry, and "the people" through a repertoire of enemy construction, platform-led mobilization, and technocratic rebranding. The ethical oil discourse relied on moral juxtaposition and adversarial politics by portraying bitumen as a commendable substitute for "theocratic" or "gangster" oil, while denouncing critics as opportunists and conspirators (Kinder, 2024; Laurie, 2019). This move normalized a friend–enemy grammar, later amplified across social media, juxtaposing a patriotic "self" against threatening "others," and helped cement a "common sense" of oil sands as a public good (Gunster et al., 2021; Neubauer et al., 2023).

The Public Inquiry into Anti-Alberta Energy Campaigns helped to crystallize the binary construction. While acknowledging that civil-society campaigns were lawful and an exercise of free expression, the Inquiry Report nonetheless deemed Alberta as under siege from a widespread network of anti-oil sands initiatives, highlighting the danger posed by a continuous, externally driven opposition (Allan, 2021). As noted in scholarly analyses of Canadian petro-politics (e.g., Chen, 2019; Gunster & Saurette, 2014; Kinder, 2024; Kuteleva & Leifso, 2020), oil sands proponents have long contended that the principal challenge confronting the Canadian oil and gas sector is communicative—a lack of "facts" and convincing storylines—and advocated for the establishment of "subsidized publics" and connective leadership to enhance populist narratives depicting the oil sands as a public good under threat.

The public inquiry translated this strategy into a state-sponsored form: it highlighted foreign funding volumes and urged the state–industry bloc to professionalize a national narrative to level the playing field. The proposed solutions were based not on regulatory reform but on communications-governmentality: a long-term branding strategy aimed at establishing Canada as a global leader in energy production and export, an educational framework for energy literacy, and a coalition structure to reduce polarization. Such solutions represented a repertoire that shifts political division to issues of knowledge, civility, and coordination, rather than material disputes regarding land, emissions, and risk. In short, the public inquiry exemplified neoliberal populism: the collective "we" is rhetorically exalted, while the operative instruments are measurement, branding, and inter-elite convening aimed at competitiveness (Matsen et al., 2016; Rossi, 2025; Scheiring, 2022).

Theoretically, the chapter has demonstrated that Canadian extractive populism retools Polanyi's (2001 [1957]) double movement. Protection is not framed as social limits on markets; it is recast as protection of markets—pipelines, investment climates, and export corridors—against cosmopolitan elites and unruly activists, while sidelining the environmental and Indigenous justice claims that once animated countermovements. Extractive populism also advances neoliberal populism's paradox: the collective is celebrated linguistically with terms such as "we" and "Canadians," while policy solutions intensify market logics and energy citizenship (Wood, 2018). Moreover, the shift from comparing ethical and unethical oil to permanent mobilization marks a radicalization of campaign style: enemy construction hardens, affective registers sharpen, and the preferred instruments are partisan-administrative (war rooms, inquiries, campaigns) rather than deliberative politics (Neubauer, 2018).

In conclusion, Canada's petro-bloc is radicalizing by using extractive populism to sustain a permanent campaign for oil sands development. This political project moralizes markets, demonizes dissent, and operationalizes "the people" through platform infrastructures and state-adjacent branding, even while acknowledging the legality of its opponents' claims. As evidenced by the Inquiry Report's recommendations, the future of the oil sands industry, according to the petro-bloc, will not involve structural decarbonization or stringent extraction limitations; rather, it will entail a managed politics of consent, wherein communicative tools mitigate conflict and prolong the existence of petro-modernity.

References

Allan, J. S. (2021). *Report of the PublicInquiry into Anti-Alberta energy campaigns*. Government of Alberta. https://open.alberta.ca/publications/public-inquiry-into-anti-alberta-energy-campaigns-report

Ambrose, J. (2020, April 20). Oil prices dip below zero as producers forced to pay to dispose of excess. *The Guardian*. https://www.theguardian.com/world/2020/apr/20/oil-prices-sink-to-20-year-low-as-un-sounds-alarm-on-to-covid-19-relief-fund

Anderson, D. (2019, December 11). Alberta's energy 'war room' launches in Calgary. *CBC News*. https://www.cbc.ca/news/canada/calgary/alberta-war-room-launch-calgary-1.5392371

Bekmen, A., & Alp Özden, B. (2022). The rise and demise of neoliberal populism as a hegemonic project: Brazil, Thailand, and Turkey. *Rethinking Marxism, 34*(3), 338–360. https://doi.org/10.1080/08935696.2022.2069459

Bellefontaine, M. (2021, October 21). Foreign funding played role in anti-resource campaigns, but Allan inquiry can't say how much. *CBC News*. https://www.cbc.ca/news/canada/edmonton/anti-alberta-energy-campaign-public-inquiry-final-report-1.6218861

Bosworth, K. (2020). The people's climate march: Environmental populism as political genre. *Political Geography, 83*, Article 102281. https://doi.org/10.1016/j.polgeo.2020.102281

Canadian Press. (2021, April 30). Failed fitting caused 190,000-litre spill at Trans Mountain site in Abbotsford, B.C. TSB. *Global News*. https://globalnews.ca/news/7755466/trans-mountain-spill-abbotsford-bc-report-tsb/

Canadian Press. (2024, January 30). A timeline of the Trans Mountain pipeline expansion's major milestones and setbacks. *Financial Post*. https://financialpost.com/pmn/business-pmn/a-timeline-of-the-trans-mountain-pipeline-expansions-major-milestones-and-setbacks

Cecco, L. (2021, March 15). Canadian lobbyists attack Netflix children's film for 'anti-oil propaganda.' *The Guardian*. https://www.theguardian.com/world/2021/mar/15/canada-lobby-group-bigfoot-family-oil-and-gas

Chen, S. (2019). From persuasion to manipulation: Tracing oil sands narratives in the *Calgary Herald*. In H. Greaves & D. Beard (Eds.), *The rhetoric of oil in the twenty-first century* (pp. 255–270). Routledge.

Clancy, C. (2019, September 09). Public inquiry into international campaigns related to energy launches website. *Edmonton Journal*. https://edmontonjournal.com/news/politics/public-inquiry-into-international-campaigns-related-to-energy-launches-website

Ecojustice. (n.d.). Trans Mountain 2.0: Challenging the federal government's project approval. https://ecojustice.ca/file/trans-mountain-2-0-challenging-the-federal-governments-project-approval

Gerbaudo, P. (2018). Social media and populism: An elective affinity? *Media, Culture & Society, 40*(5), 745–753. https://doi.org/10.1177/0163443718772192

Guardino, M. (2018). Neoliberal populism as hegemony: A historical-ideological analysis of US economic policy discourse. *Critical Discourse Studies, 15*(5), 444–462. https://doi.org/10.1080/17405904.2018.1442361

Gunster, S., Neubauer, R., Bermingham, J., & Massie, A. (2021). "Our oil": Extractive populism in Canadian social media. In W. Carroll (Ed.), *Regime of obstruction: How corporate power blocks energy democracy* (pp. 197–224). Athabasca University Press.

Gunster, S., & Saurette, P. (2014). Storylines in the sands: News, narrative, and ideology in the *Calgary Herald*. *Canadian Journal of Communication, 39*(3), 1–27. https://doi.org/10.22230/cjc.2014v39n3a2830

Hess, D. J. (2023). Pipeline conflicts, coalitions, and strategic action: A review of the literature. *The Extractive Industries and Society, 16*, Article 101339. https://doi.org/10.1016/j.exis.2023.101339

Hoberg, G. (2021). *The resistance dilemma: Place-based movements and the climate crisis*. The MIT Press.

Kinder, J. B. (2024). *Petroturfing : Refining Canadian oil through social media*. University of Minnesota Press.
Klaszus, J. (2019, December 21). Jason Kenney's year of attack. *The Sprawl*. https://www.sprawlcalgary.com/jason-kenneys-year-of-attack
Kojola, E. (2019). Bringing back the mines and a way of life: Populism and the politics of extraction. *Annals of the American Association of Geographers, 109*(2), 371–381. https://doi.org/10.1080/24694452.2018.1506695
Konings, M. (2012). Imagined double movements: Progressive thought and the specter of neoliberal populism. *Globalizations, 9*(4), 609–622. https://doi.org/10.1080/14747731.2012.699939
Kuteleva, A., & Leifso, J. (2020). Contested crude: Multiscalar identities, conflicting discourses, and narratives of oil production in Canada. *Energy Research & Social Science, 70*, Article 101672. https://doi.org/10.1016/j.erss.2020.101672
Laurie, R. (2019). Still ethical oil: Framing the Alberta oil sands. In H. Greaves & D. Beard (Eds.), *The rhetoric of oil in the twenty-first century* (pp. 169–188). Routledge.
Lim, S. H. (2023). Look up rather than down: Karl Polanyi's fascism and radical right-wing 'populism. *Current Sociology, 71*(3), 526–544. https://doi.org/10.1177/00113921211015715
Marquardt, J., & Lederer, M. (2022). Politicizing climate change in times of populism: An introduction. *Environmental Politics, 31*(5), 735–754. https://doi.org/10.1080/09644016.2022.2083478
Matsen, E., Natvik, G. J., & Torvik, R. (2016). Petro populism. *Journal of Development Economics, 118*, 1–12. https://doi.org/10.1016/j.jdeveco.2015.08.010
McCurdy, P. (2023). The rise and fall of the Synthetic: The mediatization of Canada's oil sands. *International Journal of Cultural Studies, 26*(4), 427–444. https://doi.org/10.1177/13678779231159697
Neubauer, R. (2018). Moving beyond the petrostate: Northern Gateway, extractivism, and the Canadian petrobloc. *Studies in Political Economy, 99*(3), 246–267. https://doi.org/10.1080/07078552.2018.1536369
Neubauer, R., Graham, N., & Krobath, H. (2023). Defending "Canadian energy": Connective leadership and extractive populism on Canadian Facebook. *Environmental Communication, 17*(6), 634–652. https://doi.org/10.1080/17524032.2023.2235919
Neubauer, R., & Gunster, S. (2019). Enemies at the Gateway: Regional populist discourse and the fight against oil pipelines on Canada's West Coast. *Frontiers in Communication, 4*, Article fcomm.2019.00061. https://doi.org/10.3389/fcomm.2019.00061
Polanyi, K. (2001 [1957]). *The great transformation: The political and economic origins of our time*. Boston: Beacon Press.

Queiroz, R. M. da C. (2021). From the tyranny of the neoliberal individual to neoliberal populism. *Journal of Political Ideologies, 26*(3), 240–261. https://doi.org/10.1080/13569317.2021.1885589

Ofstehage, A., Wolford, W., & Borras, S. M. (2022). Contemporary Populism and the Environment. *Annual Review of Environment and Resources, 47*(1), 671–696. https://doi.org/10.1146/annurev-environ-012220-124635

Riley, S. J. (2020, March 20). Alberta's record-low oil prices: what the coronavirus and a supply glut mean for the province. *The Narwhal.* https://thenarwhal.ca/albertas-record-low-oil-prices-what-the-coronavirus-and-a-supply-glut-mean-for-the-province/

Riofrancos, T. (2020). Resource radicals: From petro-nationalism to post-extractivism in Ecuador. *Duke University Press.* https://doi.org/10.1515/9781478012122

Rossi, N. (2025). Populism without a people: Neoliberal populism and the rise of the Italian far right. *Journal of Political Ideologies, 30*(2), 455–475. https://doi.org/10.1080/13569317.2023.2196241

Scheiring, G. (2022). The national-populist mutation of neoliberalism in dependent economies: The case of Viktor Orbán's Hungary. *Socio-Economic Review, 20*(4), 1597–1623. https://doi.org/10.1093/ser/mwac007

Sconfienza, U. M. (2022). What does "environmental populism" mean? Mapping the existing debate and a plea for dropping this label. *Populism, 5*(2), 209–235. https://doi.org/10.1163/25888072-bja10035

Thurton, D. (2020, December 6). Trans Mountain's COVID-19 costs are now $12.5 million and growing. *CBC News.* https://www.cbc.ca/news/politics/trans-mountain-pipeline-covid-pandemic-1.5829486

Tian, W., & Ge, J. (2025). Fueling divisions: A multimodal analysis of Canadian petro-nationalism in the social media discourse of "oil sands strong." *Social Semiotics, 35*(1), 70–90. https://doi.org/10.1080/10350330.2023.2261853

Wood, T. (2018). Energy's citizens: The making of a Canadian petro-public. *Canadian Journal of Communication, 43*(1), 75–92. https://doi.org/10.22230/cjc.2017v43n1a3312

CHAPTER 5

Conclusion

Abstract Synthesizing the cases, the conclusion chapter identifies recurring pitfalls of pro-bitumen discourse—affective domestication of hydrocarbons, symbolic nationalization, and populist vilification—each of which diverts attention from emissions intensity, uneven risks, and justice claims. It proposes discursive interventions that foreground transparency and lived effects and argues that "just transition" can anchor broad coalitions by linking care, dignity, and reliability to concrete decarbonization pathways. The chapter closes by outlining research agendas on platform-mediated petro-publics, ESG-driven greenwashing, and connective leadership in transition movements.

Keywords Coalition building · Discursive interventions · Energy democracy · Just transition · Polarization mitigation · Transparency

Across the three public campaigns analysed, a common architecture of persuasion becomes visible. First, lifestyle advertising displaces "oil sands" with the universal, affectively warm signifier "energy," saturating it with scenes of family intimacy, care, and everyday convenience. The "Life Takes Energy" campaign epitomized this shift: by borrowing the

S. Chen, *Petro-Rhetoric in Canada*, Palgrave Studies in Media and Environmental Communication,
https://doi.org/10.1007/978-3-032-12964-2_5

visual grammar of food and domesticity, it naturalized fossil-fuelled prosperity while backgrounding extraction, pipelines, and Indigenous jurisdiction. In discourse-historical terms, the topoi of necessity ("life requires energy") and consequence ("if energy flows, good things happen") anchored a depoliticized common sense in which carbon dependence appears benign and inevitable (Cozen, 2010; McCurdy, 2018; Thomlison, 2019).

Second, when lifestyle appeals alone are insufficient, proponents lean on symbolic nation-building and resource nationalism. In response to the political struggles over the Trans Mountain Pipeline Expansion project, the "Keep Canada Working" campaign advanced a discourse of "symbolic nationalization," suturing private sector interests to a homogenized Canadian identity and downplaying foreign ownership and financialization in the sector. This homogenizing "we"—central to resource nationalism—promises collective uplift while accusing critics of being elitist or disloyal (Barney, 2017; Conversi, 2020; Gunster et al., 2021a).

Third, as contention escalates, the petro-bloc becomes more radicalized: opponents are labelled "foreign-funded extremists," and dissent comes under increased scrutiny. This was the logic of extractive populism charted in Alberta's post-2019 political context and in the "Public Inquiry into Anti-Alberta Energy Campaigns," in which legitimacy challenges were presented as literacy, communication, and coordination problems solvable by public relations rather than democratic deliberation (Gunster et al., 2021b; Kinder, 2024; Matsen et al., 2016).

The above campaigns did not operate in isolation. They were scaffolded by a broader petro-bloc: a decentralized, interlocked coalition of state agencies, firms, financiers, media, and advocacy organizations that routinizes these discourses and supplies "inter-field" legitimacy (Neubauer, 2018). Think tanks and advocacy groups construct frames that travel easily across newsrooms, legislatures, and social platforms, thereby facilitating the formation of a subsidized petro-public. Furthermore, the establishment and perpetuation of the petro-bloc are facilitated by Canada's media structure dynamics: concentrated ownership, pro-industry editorial perspectives, and the minimization of climate issues in the coverage of mega-projects all normalize the terms of debate (e.g., employment, economic expansion, and nation-building) while sidelining environmental considerations (Carroll, 2021).

Overall, Canadian pro-bitumen discourse since the mid-2010s has evolved from technocratic reassurance to a two-track strategy: (1) domesticating carbon through lifestyle imagery and abstract "energy" talk, and (2) polarizing politics via extractive-populist appeals that morally elevate extraction and stigmatize opponents. This evolution entrenches petroculture as common sense, impedes decarbonization, and reframes democratic contestation as a threat to prosperity and national unity.

5.1 The Pitfalls of Pro-Oil Sands Discourse

Pro-bitumen advocacy in Canada has not merely defended a sector; it has constructed an everyday myth that naturalizes carbon dependence and narrows the range of possible energy futures. Debunking this myth can begin from the following perspectives. First, climate denial has become denial by displacement. Pro-bitumen messaging rarely contests the science; it redirects public attention to life empowered by energy, obscuring greenhouse gas emissions and other environmental harms. Everyday life is consequently rhetorically intertwined with fossil fuel consumption, while planetary constraints are relegated to the periphery as someone else's problem, in another place, at a later time. To counter this, it is imperative to continue informing the public about the substantial harm already inflicted by the climate emergency (McHugh et al., 2021). The sustained public attention to wildfires, for instance, offers a useful starting point for initiating discussions regarding decarbonization and climate change mitigation (Hopke, 2020).

Second, the promise of shared prosperity from oil sands development sits uneasily with distributional realities. Populist appeals pledge collective uplift while obscuring fiscal flows, rents, and the socialization of risk (Kraushaar-Friesen & Busch, 2020; Riofrancos, 2020). When public money underwrites private mega-projects or acquisition risk, the discourse of national solidarity masks asymmetries of gain and loss. Media frames amplify this by privileging jobs and growth talking points and casting environmental scrutiny as hostility to the nation. Accordingly, additional, impartial assessments of the economic benefits and risks associated with the development of oil sands are urgently needed (Chen, 2023).

Third, besides climate delay and technocratic frames, Canada's environmental communication is replete with greenwashing. For example, the Pathways Alliance, a coalition of the six largest oil sands producers, has launched high-profile public campaigns pledging future emissions reduction (asserting goals like "net zero by 2050") while downplaying projected near-term production growth (Aronczyk et al., 2024). In response, regulatory and public interest oversight needs to catch up and establish explicit guidelines concerning greenwashing advertisements. Recently, the Competition Bureau opened an investigation into potentially misleading environmental representations, and, amid tightening anti-greenwashing rules, the Pathways Alliance removed environment-related claims from its website and social media channels (Nickel, 2023).

Last but not least, pro-oil sands discourse reframes critical questions concerning Indigenous sovereignty into communications problems. Frames that homogenize Canadian collective identity subordinate Indigenous laws and titles to stakeholder perspectives, despite judicial findings of inadequate consultation or unmet legal obligations. This rhetorical omission is not accidental; it is fundamental to a politics that treats legitimacy as a matter of persuasion rather than consent. This situation calls for further coalition building efforts among environmental groups, concerned settlers, and Indigenous communities.

5.2 Mitigating Polarization

Polarization in Canadian energy debates is sustained not just by antagonism but also by structural and rhetorical factors. Accordingly, interventions must blend policy design with rhetorical craft. To begin with, security and prosperity ought to be redefined as resilience rather than mere GDP growth. One important lesson from the "Life Takes Energy" campaign is not to imitate its mystifications but to compete on the terrain of everyday life. Thus, security and prosperity can be discussed not only in terms of reliable, affordable heat and mobility, but also in terms of clean air, disaster-ready communities, and sustainable employment opportunities. Environmental communications ought to associate these issues with concrete, near-term programmes—such as extensive retrofits, decentralized storage, increased transit frequency, and grid modernization—whose benefits are realized more swiftly than the returns from any new pipeline.

Furthermore, as populist storylines flourish in contexts where costs and benefits are obscured, technocracy ought to be replaced by transparency. For example, municipal "energy accounts"—public dashboards that monitor project-level emissions, public subsidies, job quality, and household bill effects—can challenge the narratives surrounding the inevitability of carbon dependency and ground discussions in lived experience rather than abstract talking points (Berthod et al., 2023; Szulecki & Overland, 2020).

Third, considering the petro-bloc's advantages in subsidized publics and connective leadership—ranging from think-tank reports and op-eds that resonate as common sense to social media accounts that aggregate sentiment on a large scale—the communication sector must be democratized to counteract petro-hegemony. Strategies including public interest media funding, regulations mandating platform transparency regarding environmental claims, and connective leadership in movement-oriented transition coalitions could be employed to cultivate frameworks of care, dignity, and reliability, rather than solely addressing misinformation (Hackett et al., 2017).

Fourth, for coalition building, the concept of "just transition" can serve as a rallying point for cross-sector collaboration on social change. By integrating climate, energy, and environmental justice, the concept offers a shared language of guarantees—decent work, equity, and social dialogue—that unions, Indigenous communities, municipalities, and environmental groups can co-own and operationalize through concrete instruments such as community-benefit agreements and co-ownership models (Hess, 2018). Moreover, the concept aligns with the energy democracy literature and multi-coalition strategy research, which show how cross-field alliances and connective leadership can scale participation and maintain message discipline without collapsing pluralism—key to reframing conflicts away from the "jobs versus the environment" binary (Szulecki & Overland, 2020).

5.3 Future Research Directions

This book's main findings suggest four directions for future research. First, platformed petro-publics necessitate detailed, mixed-methods analysis. Future research may concentrate on network cartographies of Canadian pro-fossil ecosystems on Facebook, YouTube, and TikTok; archives of A/B-tested message variants; and ethnographies of connective leadership. The goal is to further assess the impact of "petroturfing" (Kinder, 2024) as durable mobilization infrastructure and to explore feasible counter-infrastructure for coalition building among labour, Indigenous, municipal, and climate-justice stakeholders.

Second, the prevalence of greenwashing and climate delay narratives calls for longitudinal reception studies that examine which disclosures—such as project-level emissions, abatement timelines, and public subsidy amounts—actually change perceptions of net-zero claims. Experimental designs and field trials can further assess which environmental education materials reduce susceptibility to deceptive greenwashing messaging. Building on petroculture scholarship (e.g., Wilson et al., 2017), researchers can also test the effectiveness of narrative frames (care-first, reliability-first, anti-monopoly) that connect everyday life (e.g., lower bills, quieter streets, cleaner air) to structural decarbonization policies across regions and social classes.

Third, inquiries regarding energy democracy, social license, and political legitimacy should be further integrated into communication research. Such inquiries may include, but are not limited to: How do the ownership stakes of First Nations, municipal utilities, and co-ops affect the narrative of consent for energy infrastructure? Will communities be more inclined to grant "social license" when they are also the owners of renewable energy projects?

Fourth, populism beyond the right warrants ongoing academic attention. Has Canadian energy discourse been influenced by extractive populism, and can a left-populist framework—focused on dignity, public goods, anti-monopoly, and community sovereignty—engage diverse constituencies without creating reciprocal polarization (Gunster et al., 2021a)? Identifying antagonisms that broaden rather than limit policy imagination is a priority.

In sum, pro-oil sands narratives in Canada have been remarkably effective at symbolically nationalizing oil sands projects and presenting support for carbon dependency as common sense. The counter-task is to turn common sense back into discussions about humanity's energy future. This requires coupling legal and policy change with narrative strategies that redefine prosperity and security, make costs and benefits legible, honour Indigenous sovereignty, and tell transition as stories about happy and sustainable everyday life.

References

Aronczyk, M., McCurdy, P., & Russill, C. (2024). Greenwashing, net-zero, and the oil sands in Canada: The case of Pathways Alliance. *Energy Research &*

Social Science, 112, Article 103502. https://doi.org/10.1016/j.erss.2024.103502

Barney, D. (2017). Who we are and what we do: Canada as a pipeline nation. In S. Wilson, A. Carlson, & I. Szeman (Eds.), *Petrocultures: Oil, politics, culture* (pp. 86–126). McGill-Queen's University Press.

Berthod, O., Blanchet, T., Busch, H., Kunze, C., Nolden, C., & Wenderlich, M. (2023). The rise and fall of energy democracy: 5 cases of collaborative governance in energy systems. *Environmental Management, 71*(3), 551–564. https://doi.org/10.1007/s00267-022-01687-8

Carroll, W. K. (ed.) (2021). *Regime of obstruction: How corporate power blocks energy democracy*. Athabasca University Press.

Chen, S. (2023). The fate of bitumen: an exploratory study of national newspaper coverage of Alberta's bitumen industry during the COVID-19 pandemic. *Humanities & Social Sciences Communications, 10*, Article 30. https://doi.org/10.1057/s41599-023-01515-2

Conversi, D. (2020). The ultimate challenge: Nationalism and climate change. *Nationalities Papers, 48*(4), 625–636. https://doi.org/10.1017/nps.2020.18

Cozen, B. (2010). This pear is a rhetorical tool: Food imagery in energy company advertising. *Environmental Communication, 4*(3), 355–370. https://doi.org/10.1080/17524032.2010.499212

Gunster, S., Fleet, D., & Neubauer, R. (2021a). Challenging petro-nationalism: Another Canada is possible? *Journal of Canadian Studies, 55*(1), 57–87. https://doi.org/10.3138/jcs.2019-0033

Gunster, S., Neubauer, R., Bermingham, J., & Massie, A. (2021b). "Our oil": Extractive populism in Canadian social media. In W. Carroll (Ed.), *Regime of obstruction: How corporate power blocks energy democracy* (pp. 197–224). Athabasca University Press.

Hackett, R. A., Forde, S., Gunster, S., & Foxwell-Norton, K. (Eds). (2017). *Journalism and climate crisis: public engagement, media alternatives*. Routledge.

Hess, D. J. (2018). Energy democracy and social movements: A multi-coalition perspective on the politics of sustainability transitions. *Energy Research & Social Science, 40*, 177–189. https://doi.org/10.1016/j.erss.2018.01.003

Hopke, J. E. (2020). Connecting Extreme Heat Events to Climate Change: Media Coverage of Heat Waves and Wildfires. *Environmental Communication, 14*(4), 492–508. https://doi.org/10.1080/17524032.2019.1687537

Kinder, J. B. (2024). *Petroturfing: Refining Canadian oil through social media*. University of Minnesota Press.

Kraushaar-Friesen, N., & Busch, H. (2020). Of pipe dreams and fossil fools: Advancing Canadian fossil fuel hegemony through the Trans Mountain

pipeline. *Energy Research & Social Science, 69*, Article 101695. https://doi.org/10.1016/j.erss.2020.101695
Matsen, E., Natvik, G. J., & Torvik, R. (2016). Petro populism. *Journal of Development Economics, 118*, 1–12. https://doi.org/10.1016/j.jdeveco.2015.08.010
McCurdy, P. (2018). From the natural to the manmade environment: The shifting advertising practices of Canada's oil sands industry. *Canadian Journal of Communication, 43*(1), 33–52. https://doi.org/10.22230/cjc.2018v43n1a3315
McHugh, L. H., Lemos, M. C., & Morrison, T. H. (2021). Risk? crisis? emergency? implications of the new climate emergency framing for governance and policy. *Wiley Interdisciplinary Reviews. Climate Change, 12*(6), Article e736. https://doi.org/10.1002/wcc.736
Neubauer, R. (2018). Moving beyond the petrostate: Northern Gateway, extractivism, and the Canadian petrobloc. *Studies in Political Economy, 99*(3), 246–267. https://doi.org/10.1080/07078552.2018.1536369
Nickel, R. (2023, May 11). Canada's Competition Bureau investigates oil sands group over advertising. *Reuters*. https://www.reuters.com/world/americas/canadas-competition-bureau-investigates-oil-sands-group-over-advertising-2023-05-11
Riofrancos, T. (2020). Resource radicals: From petro-nationalism to post-extractivism in Ecuador. *Duke University Press*. https://doi.org/10.1515/9781478012122
Szulecki, K., & Overland, I. (2020). Energy democracy as a process, an outcome and a goal: A conceptual review. *Energy Research & Social Science, 69*, Article 101768. https://doi.org/10.1016/j.erss.2020.101768
Thomlison, A. (2019). From pipeline to plate: The domestication of oil sands through visual food analogies. In H. Greaves & D. Beard (Eds.), *The rhetoric of oil in the twenty-first century: Government, corporate, and activist discourses* (pp. 189–210). Routledge.
Wilson, S., Carlson, A., & Szeman, I. (Eds.). (2017). *Petrocultures: oil, politics, culture* (1st ed.). McGill-Queen's University Press. https://doi.org/10.1515/9780773550391

Index

S. Chen, *Petro-Rhetoric in Canada*, Palgrave Studies in Media and Environmental Communication,
https://doi.org/10.1007/978-3-032-12964-2

FSC
www.fsc.org
MIX
Papier | Fördert
gute Waldnutzung
FSC® C083411

Kolibri 360 GmbH
Ferdinand-Jühlke-Straße 7
99095 Erfurt, Deutschland
produktsicherheit@kolibri360.de